Making Profits with Dairy Cows and Quotas

Making Profits with Dairy Cows and Quotas

GORDON THROUP

FARMING PRESS

First published 1994

ISBN 0 85236 281 1

A catalogue record for this book is available
from the British Library

Published by Farming Press Books
Wharfedale Road, Ipswich IP1 4LG, United Kingdom

Distributed in North America
by Diamond Farm Enterprises,
Box 537, Alexandria Bay, NY 13607, USA

Cover photograph courtesy of Genus Marketing

Cover design by Andrew Thistlethwaite
Typeset by Galleon Typesetting, Ipswich
Printed and bound in Great Britain by
Butler and Tanner Ltd, Frome and London

Contents

Preface

The introduction of milk quotas in 1984, based on the actual production achieved in the year ended 31 December 1983, has had a dramatic effect on the business plans and financial success achieved by farmers. The first question one asks when visiting a dairy farm these days is 'How much quota have you got?', not 'How many acres?', as this is now the most limiting resource available to the dairy farmer. On many farms a farming system has been evolved to fit the quota whereas my motto, since the introduction of quotas, has been 'Make the quota fit the farm—do not make the farm fit the quota.'

A second crucial change in the dairy farmer's environment during the past ten years has been the change in the value of silage and other alternative forages compared to their cost of production. In 1983 the cost of concentrates was in the region of £165 per tonne, and brewers' grains, depending on the time of year, were in the region of £20–25 per tonne, ie prices similar to those at the present time. During the same period the price received for milk has increased from 13p per litre to 22p per litre, giving the most favourable ratio ever between the milk price relative to feed costs. Despite this, the perceived wisdom of many farmers and their advisers has been to aim to continue to base their farming policy on producing more milk from forage. This in part explains why there has been very little increase in the national yield per cow during the past ten years despite the availability of improved genetics.

At the present time, however, the opportunist dairy farmer should be looking at ways to exploit cheap purchased feed including such items as brewers' grains instead of simply aiming to produce more milk from forage. At the same time he should be looking at ways to exploit genetics and increase the yields he achieves per cow. In other words, he has to decide whether he is a forage/grass farmer whose main objective is to use cows to produce milk from forage or whether he is a dairy farmer whose main objective is to use grass and other purchased feed to exploit the genetic/profit potential of his dairy cows. Economic circumstances at the present time certainly favour the latter but of course this was not the case in the 1970s/80s when feed prices were high relative to milk and other product prices.

Dairy farmers in the West and North West, ie in the grass-favoured growing areas, do not have the same opportunity to exploit alternative feeds as those in the Midlands and South. During the next decade, therefore, we could see the development of two trends: farmers in the West and North West concentrating on producing milk from grazed grass, as is the case at the present time in Southern Ireland and New Zealand. This could be associated with close links to processing factories that could close down in the winter months. The farmers in the Midlands and South, ie close to the main conurbations, and with the ability to grow forage crops, are likely to move towards a system aimed at high yields per cow, with an emphasis on an even supply of milk throughout the year.

Since the original base year of 1983 the total quota cuts have been in the region of 21% so any dairy farmer who has not already acquired extra quota by either leasing or buying will be operating at 80% of the 1983 level. His fixed costs per litre will have risen accordingly, hence a favourite phrase of mine: 'Today's problem in dairy farming is high fixed costs, not high feed costs.'

Farmers who were prepared to take both the risk and the opportunity purchased quota several years ago, the very brave at prices in the region of 11–12p per litre within two years of milk quotas being introduced. A few others took the opportunity to purchase milk quota in 1992/93 when its price fell below 30p per litre. The remainder will need to wait for the next opportunity as quota prices at the time of writing are in the region of 48p per litre.

This leads me to quote a phrase that has become familiar to many of my friends and colleagues since quotas were introduced:

> 'If the opportunity to make money comes along
> and you don't see it, you won't make it.
> If the opportunity comes along and you see it
> and you are not ready, you won't make it.
> If the opportunity comes along and you are not
> prepared to take the risk, you won't make it,
> and
> If everybody else sees it, don't take it!'

In other words, dairy farming has become very opportunistic in nature. Dairy farmers who are prepared to take the opportunity have already increased their production to the same level as, or preferably to more than, they were achieving in 1983 by either leasing or purchasing the required quota, preferably the former.

The role on the dairy farm of dairy heifers and other supplementary/complementary enterprises to the dairy herd has changed as a result of the introduction of quotas.

Prior to the introduction of quotas many dairy farmers were justifiably criticised for carrying too many dairy replacements. On many farms now where capital and quota supply is restricted, rearing more dairy heifers/keeping other supplementary enterprises is crucial to the success of the farming business. Despite this statement my advice to dairy farmers would still be to produce as much milk as possible and where possible remove the constraint imposed by milk quotas.

At the time of writing, dairy farming is going through one of its periods of prosperity and all the talk is about whether or not to join Milk Marque or one of the other dairies competing for farmers' supplies. In a book about farm business management one would possibly expect a large section to be devoted to the question as to whom a dairy farmer should sell his milk. In most cases this is likely to be a once-in-a-lifetime decision and most producers will have made their decision prior to this book being printed.

This book is written on the assumption that most milk producers will sign up to sell their milk to Milk Marque to the benefit of all producers, including those who do not join Milk Marque. Niche markets will certainly exist for some producers and it will pay them to join organisations other than Milk Marque and in due course producers joining Milk Marque may have to review their decision.

To return to the question of prosperity, dairy farming profits at the present time are probably higher than they have been at any time during the past 30 years, including 1972 to 1973. Financial data has to be used from time to time in the book to illustrate profitability points and one is very conscious that this can quickly become out of date and that profitability can fall.

Dairy cow prices at the present time are at a record high. A theme running through the book is that in dairy farming one has a capital asset in the form of the dairy cow as well as trading profits from milk production. Profits in dairy farming go down as well as up, and this underlines the opportunist nature of dairy farming and the need for dairy farmers to study profitability trends as well as the technical aspects of their business.

Dairy farming is still a way of life for many dairy farmers and long may it remain so. Despite the long hours it is a very satisfactory way of making a good living, providing the business is properly organised and managed, and this is the theme of this book.

GORDON THROUP
August 1994

Making Profits with Dairy Cows and Quotas

SECTION ONE

The Principles of Dairy Farm Management

CHAPTER 1

Business Management and the Dairy Farmer

THE NEED FOR A BUSINESS APPROACH

Traditionally farming has been seen and treated as a way of life. In today's competitive world this is no longer feasible but most people still go into, or stay in, farming because they like the job, not because 'the money's good'. Most students at agricultural college are asked at some point why they want to go into farming. Many answers to this question are given, such as:

'I like the open air life.'
'My grandfather was a farmer.'

Only rarely is the answer 'I'm in it for the money,' or 'It's a secure job and has good career prospects.'

Farmers' sons and daughters, however, are beginning to question whether they should simply follow in father's footsteps. They are looking at the lifestyles enjoyed by their contemporaries, and asking the question 'Do I want to work longer hours than most people appear to do, and not have greater financial rewards?'

A characteristic of dairy farming is the relatively small size of the individual business. In the past there were very few opportunities for an individual without capital to pursue a career in dairy farming. The past 20 to 30 years, however, have seen a drastic reorganisation in the size and degree of specialisation in dairy farming and this has opened up new job opportunities.

Landlords have taken in hand land previously let out to tenants, and institutional investors have also invested in farming. These new farmers differ from the traditional ones in their need for management expertise to run the farms and in their business approach to farming. The majority of British farmers, however, still differ from other businesses in the commercial world in the sense that they provide both the capital and the management expertise to run the business.

The more traditional or bona fide farmer has also had to increase the size of his business to survive and this has increased his need for specialist farmworkers. This has been particularly true in dairy farming where larger herds have resulted in the need for herd managers and herdspersons rather than large numbers of less skilled cowmen.

The increasing size of their farms has also led bona fide farmers to adopt a much more business-like approach to their farming. This need for a business-like approach has been intensified by the cost/price squeeze farming has undergone in recent years. As a result of these changes there have been enormous increases in productivity! This allows milk to leave the farm gate in the 1990s at a price which, in real terms (ie the cost of wages), is less than 50% of that received 20 years ago. See Table 1.1.

Table 1.1 Milk price compared with cowmen's wages

	*Milk price** *(pence per litre)* *England and Wales*	*Weekly earnings and hours worked by dairy cowmen in Great Britain†*		*Number of litres equivalent to weekly earnings*
Year		*Earnings £*	*Hours*	
1954–55	3.3	8.72	57.1	262
1959–60	3.2	11.64	56.9	361
1964–65	3.4	15.15	56.4	445
1969–70	3.6	20.63	54.8	581
1974–75	6.4	45.03	53.2	721
1979–80	11.6	92.42	53.3	802
1984–85	14.6	148.14	52.0	1,016
1989–90	18.9	197.87	51.1	1,044
1991–92	19.7	248.34	52.0	1,260

* MMB Dairy Facts and Figures.
† MAFF and DAFS.

Dairy farmers, like other businessmen, have had to learn to live with inflation and with violent fluctuations in profitability from year to year. Back in 1972 to 1974, the price of feed increased during an 18 month period by approximately 100% due to a world grain shortage and this gave the impetus for the development of grass forage systems based on silage, the objective being to reduce feed costs. Ten years ago we saw the introduction of milk quotas, which led to a substantial reduction in the price of dairy cows and youngstock. Quotas imposed curbs on production and farmers have now had to learn to live with milk quota as a commodity and come to terms with the need to lease in and buy, or perhaps even sell milk quota in order to retire.

Farming and business skills have been required to survive the various pressures that have been exerted on dairy farms. The number of dairy farmers has fallen steadily during the past 40 years, from 142,000 in 1955, to 47,000 in 1979 to just under 29,000 in 1994. This trend is expected to continue, resulting in no more than 20,000 dairy farmers in 10 to 15 years' time.

The need for a professional approach to the business aspects of farming has to be reconciled with the typical farmer's dislike of office or paperwork and his preference for getting on with the 'proper job' of farming. This has been neatly reconciled in many instances by giving the paperwork to the son or daughter recently returned from college, or to a qualified secretary/consultant. This keeps them happy until they realise that doing the paperwork is not the same as managing the business.

MANAGEMENT FUNCTIONS IN DAIRY FARMING

Management is basically about decision-making—that is, deciding what to do and then doing it. The decisions vary in the frequency with which they have to be made and their relative importance. Some, such as whether or not to buy a farm, may only be made once in a lifetime but others, such as how many concentrates or how much grass to feed the dairy cows, are made every day.

A characteristic of farming—and this is particularly true of dairy farming—is the number of decisions that have to be taken by key workers. Dairy farming has been mechanised and automated to a considerable extent but dairy herdspersons still have a great deal of control, or at least potential control, over how a job is done. They may also decide whether a particular job is done and, probably most important, whether a job is done in the right way at the right time.

To make the correct decision about, for example, the level of concentrate feeding to cows at differing stages in lactation, herdspersons need to understand the economic as well as the husbandry implications of the decision. Dairy herdpersons therefore, as well as dairy farmers and managers, need to be capable of making and implementing decisions based on sound business management principles.

Returning to the functions of the dairy farmer or farm manager:

What sort of jobs does he do? What does dairy farm management involve?

Firstly, it involves making decisions to determine the farm policy or strategy of the business. These decisions will determine the quantity of milk that is to be produced, how much quota should be owned/leased, which in turn will determine the number of cows on the farm, the number of men to be employed, the type of machines to be purchased and the method of milk production and systems of grassland production to be adopted. These decisions will influence how the capital available is used in the business and whether or not there is a need for borrowed capital. These decisions may be made alone, as in the case of a one-man dairy farm, or after consultation with the owner and other advisers, in the case of a manager of a 400 hectare estate farm. The extent to which the manager is involved in these decisions largely determines his authority and responsibilities in the business, and hence the influence he can exert on the overall profitability of the business. In turn this tends to determine the remuneration he can command.

A major decision that has to be taken is whether milk quota should be purchased or leased, or whether the farm policy should be based on the amount of quota that happens to be available. These policy decisions, particularly those relating to milk quota, are fundamental to the success of the farm. A large part of this book is therefore devoted to the economic and management principles these involve and to the management techniques that have been developed to allow them to be put into practice.

The second stage in management is to implement the policy, or to put the plans and policies of organisation into operation. This involves many functions and occupies most of a manager's time. It is probably fair to say that most managers spend only 5% of their time deciding what to do and 95% doing it.

It is vital to get the 5% right, ie the decision what to do; otherwise the efforts in the remaining 95% of one's time can tend to be ineffectual. This is particularly true in relation to the viability of the business plan; if the business plan is not capable of producing a good profit, then no matter how good the day-to-day management, a reasonable profit will not be achieved.

The most important area of day-to-day management in dairy farming concerns control of the factors that determine the margin over concentrate costs. This involves the control of the many husbandry factors that determine milk yield per cow and yields of forage and grassland per hectare. This control is crucial to the success of the business and is discussed at length later in the book.

On many dairy farms husbandry control is direct, that is to say the farmer or manager himself is directly involved in milking the cows and the day-to-day management of other enterprises. On the larger farm many of these day-to-day decisions become the responsibility of the enterprise manager or dairy herdsperson. Consequently the job of the manager changes in emphasis and becomes much more concerned with the control and motivation of staff. This transition from managing crops and animals to managing people is one that many farm managers and farmers who expand their businesses find difficult to make. One of the biggest problems is the difficulty they find in delegating to others the decisions they previously took, such as deciding how much concentrates to feed each cow. They no longer have at their fingertips or at the back of their mind the information they require to manage the business and they have to pay more attention to other methods of collecting this information.

This lack of contact with doing the job applies as much to the paperwork or office work as it does to the running of individual enterprises. On the small farm the bookwork takes only a few hours per month and the cash book, amongst other things, is written up by the farmer. On the larger farm this is not feasible and administration or control of the farm office becomes an important aspect of the manager's job.

On larger farms the term 'administrator' describes the job of the manager more accurately than the term 'farm manager' and it is fair to say that working farm managers promoted to such a position sometimes find they enjoy less job satisfaction. It needs to be stressed that one of the arts of management on the larger farm is the ability to control a job without doing it yourself. Managers often find they can retain control by doing a small part of it themselves, for instance by adding and checking the totals in the cash analysis book completed by the secretary or by spending half an hour each week with the herd manager and/or herdsperson checking lactation trends and planned feeding levels.

Communication and the need for it both up and down the lines of authority become increasingly important the greater the size of the farm. Fortunately in farming there are few large businesses by industrial standards and effective communications can be achieved without formal procedures. Nonetheless it remains an important function of the senior management or farm manager to see that policy decisions are communicated to all staff and acted upon. Equally it is incumbent upon middle management or enterprise managers/herdspersons to inform the farm manager of the outcome of these decisions. Many decisions have to be taken with uncertain knowledge of their outcome and time or experience may prove them to be wrong. For example, a decision not to feed concentrates in summer may be a policy decision on certain farms and justified in most cases but means need to be found whereby this decision can be changed when necessary, such as in the drought conditions of 1990 and 1991.

Buying and selling is an area of management that gives a lot of job satisfaction to farmers and managers. It is sometimes said that farmers stay in agriculture simply because it gives them the opportunity to make their weekly visits to the local market. However, the professional manager regards this as largely wasted time and much prefers to make the decisions by telephone.

Marketing is the 'in word' in farming these days. Farmers and managers are urged to spend more time marketing their produce. In most dairy farming business, however, the scope for developing marketing skills is limited. There are obviously exceptions, such as producers who are also cheesemakers, and this applies to any producer who decides to go down the milk producer/processor route.

Most dairy farmers in the short term, however, are likely to take a decision to sell their milk to either Milk Marque or one of the other large dairies, and delegate the job of selling to them. It is up to each farmer/manager to try to ensure that after this has been delegated the job is effectively carried out.

Having delegated the marketing of the milk, one must seek to market the other products on the farm as effectively as possible, that is calves, cull cows and, in particular, dairy heifers and dairy cows sold surplus to requirements.

Marketing also becomes very important if a decision is taken to diversify into some form of business outside the farm.

SPECIALISATION IN MANAGEMENT

The fact that most farm businesses are small in industrial terms has already been mentioned, and this means that there is relatively little specialisation in management. One rarely finds financial managers, sales managers, personnel managers, works managers, etc each with their specialist functions. On the majority of farms all these functions have to be performed by the same person and on the smallest farm—the one-man farm—the manager is also the workforce!

It has, however, to be accepted that the size of many farm businesses is growing and thought has to be given to what form of management structure to adopt. The management structure on a 400 hectare mainly dairy farm, keeping 450 dairy cows, 400 youngstock and growing 300 hectare of forage crops and 100 hectares of cereals,

Figure 1.1 Management structure on a 400 hectare mainly dairy farm

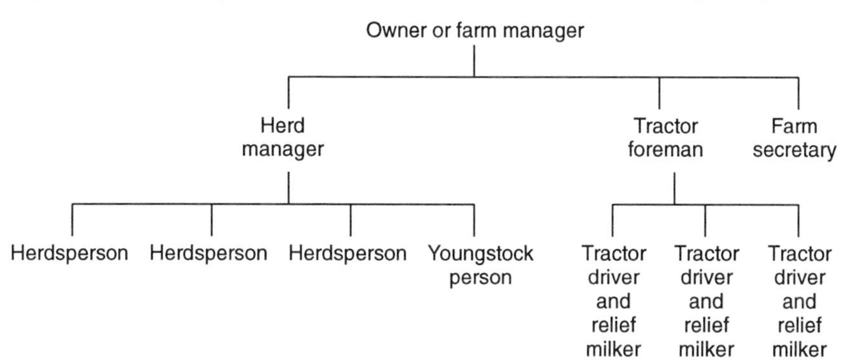

could be on the lines shown in Figure 1.1. On such a farm it is necessary to be clear in defining the responsibilities and authority that goes with each individual job so that each person knows what is expected from him or her.

In Figure 1.1 it is assumed that all the tractor drivers act as relief milkers but on many farms this would not be feasible. Consequently the herd manager could find that one of his/her main jobs is to act as a relief milker to the herdspersons under his/her control. This will probably be necessary anyway if he or she is to be gainfully employed as there is not sufficient 'management work' to justify both a herd manager and a farm manager's employment. This will depend, however, to some extent on the nature and inclinations of the manager as well as the capabilities of the tractor foreman and whether the latter is simply a foreman or an arable manager capable of taking more responsibility for decisions such as which spray to use as well as when to apply it. Given the above management structure there will be very little management responsibility for the herdsperson and this title is a misnomer. It would be more accurate to use the rather more old-fashioned terms 'cowmen' or 'cowgirls' because most of the time they would be carrying out the work under close supervision.

Most hands-on farm managers would find the management structure set out in Figure 1.1 rather irksome and, given today's farming economic climate, would be aiming to reduce the number of staff. The herd manager's role would tend to be merged with one of the herdspersons and he or she would probably be given the title 'head herdsperson' and would be responsible for the care of the cows when the farm manager was away. The same is true of the tractor foreman. Four tractor staff are shown in this chart, but ideally this would be reduced to three, and in today's climate, possibly two, with some of the work contracted out. This re-emphasises the point made earlier that on a dairy farm, even a 400 hectare farm, there is a danger of there being too many people who want to manage the farm.

JOB DESCRIPTIONS

Given the above structure, the job description and management responsibilities of the farm manager and head herdsperson would be along the lines shown below:

Farm manager

He is responsible for the implementation of the farm business policy and strategy agreed at regular meetings with the owner and his advisers, and he is accountable directly to the owner. He is required to liaise closely with the estate land agent and estate foreman.

The duties of the farm manager include:

(a) The preparation for approval annually of a profitable working, cropping, stocking and capital programme together with appropriate financial budgets.
(b) The purchase and sale of such goods and services as are required to implement the agreed policy.
(c) The day-to-day control and organisation of farm staff. Responsibility for the recruitment and dismissal of farm staff according to policy and after consultation with the owner.
(d) The day-to-day control of all farming operations in accordance with the agreed policy.
(e) The production and keeping of such records as may be required for the profitable functioning of the farm.
(f) The keeping of farm machinery and equipment and the maintenance of farm buildings and premises in a tidy condition.
(g) The supervision of capital improvements as and when agreed as part of the policy.

Head herdsperson

The head herdsperson should:

(a) Prepare annually for approval a dairy herd farming programme including proposed culling and replacement policies, gross margin budgets for each individual herd and capital expenditure proposals.

(b) Implement the approved dairy herd farming programme under the overall control and supervision of the farm manager.
(c) Accept responsibility for keeping all records pertaining to the dairy herd.
(d) Supervise all staff employed on a permanent or temporary basis to look after the dairy cows and youngstock.
(e) Liaise closely with the tractor foreman in all matters relating to the management of the dairy herd and grassland.

COST CENTRES

Substantial dairy enterprises are often found on mixed arable/livestock farms. On a 400 hectare farm, one would often find that the number of dairy cows was not more than 300, with approximately half of the land being devoted to cereals and other combine crops. One person would be mainly responsible for the arable part of the business and another for the livestock section.

On farms of this size considerable improvement in efficiency can often be achieved if two businesses are set up, one arable and one dairy, each with its own separate set of accounts. This leads quickly to the establishment of true costs of production as the arable unit has to charge the dairy unit for such items as the contract-making of silage.

The management structure would be essentially the same as that shown in Figure 1.1, but the tractor foreman would become arable manager. He would probably find that he had difficulty in justifying a staff of two and would seek to do outside contract work in

Figure 1.2 Management structure on a 400 hectare farm, with 300 dairy cows, 250 youngstock, 200 hectares forage and 200 hectares combine crop

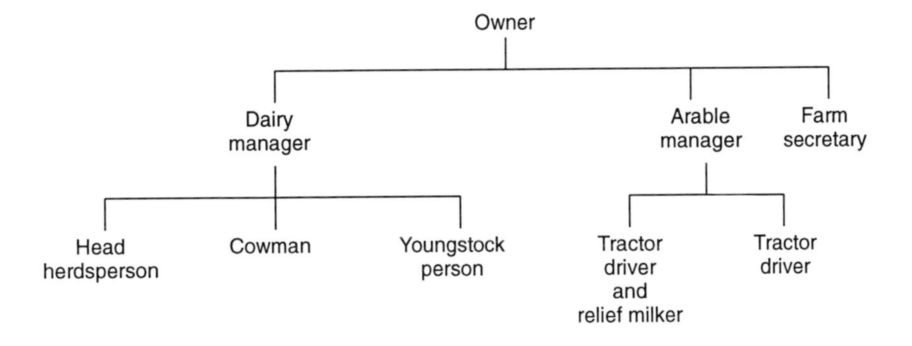

order to retain his staff, or alternatively have to accept that redundancies must be made. The management structure would be as shown in Figure 1.2.

MANAGEMENT OBJECTIVES

An agricultural economist would define the main objective as being to maximise the profit that could be obtained from the business. In practice however there are many differing objectives to satisfy, some of which conflict with the maximisation of profit.

The most important factor to consider is one's own personal objective in relation to the business in which one is working. This means asking oneself three questions and these are the same whether you are a herdsperson, manager or owner-manager:

1. What am I trying to do?
2. What is stopping me from doing it?
3. What can I do about it?

These are essential to the job of management whether you are considering the long-term future of your business, trying to decide what to do next week or determining the goals in your life. The purpose of this book is to help you ask and answer these questions because fundamentally that is what management, and life, are all about.

CHAPTER 2

The Economics of Dairy Farm Management

BASIC ECONOMIC OBJECTIVES

The basic economic objective in dairy farming, as in the management of any farm, is to combine land, labour and capital in such a way as to maximise profit, or at least to obtain a substantial profit over the long term.

To achieve this economic objective the farmer or manager has first to decide what to do and second to do it. This chapter is concerned primarily with the economic principles underlying the decision of what to do, or to put it more professionally, deciding on the overall plan and objectives of the business. To arrive at the farm policy, management has to decide:

1. (a) Which enterprises to have on the farm
   (b) How large each enterprise should be
2. (a) The method of production to adopt for each enterprise
   (b) The yield at which to aim

The basic economic objective is to maximise the profit from the whole farm, not just from the dairy herd. The economic principles involved in these decisions relate mainly to the degree of specialisation and intensity of production to adopt in running the farm.

Relatively few farms are devoted one hundred per cent to milk production. The objective in the first part of this chapter is to consider some of the reasons why, and at the same time to consider the economic principles and other factors that need to be taken into account when deciding on the number and size of each individual enterprise.

A second feature of dairy farming is the wide variation found in methods of production and yield levels. The second part of the chapter considers the economic and other factors to be taken into account in deciding what system of production to adopt and the yield level at which to aim.

SPECIALISATION

Having said that very few farms are devoted one hundred per cent to milk production, it is fair to say that dairy farms are among the most specialised in British agriculture.

Specialisation in milk production is feasible because dairy farming can be practised over a wide range of soil types and there are fewer technical problems to overcome compared, for example, with specialised cereal farms. A feature of farming in recent years has been the increased specialisation that has taken place in all farming systems and this has been particularly true of dairy farming. The reasons for this are numerous and the more significant are itemised below:

- The need to increase cow numbers to make effective use of improved housing, feeding and milking systems.
- The considerable increase in the technological knowledge required in all enterprises. An advantage of specialisation in dairy farming, as in other enterprises, is the opportunity it affords to come to terms with this technology and to implement it effectively.
- The need to eliminate less profitable enterprises to overcome problems generated by the cost/price squeeze.

An indication of the cost/price squeeze the dairy industry has undergone can be seen from an examination of Table 1.1 (Chapter 1).

1,260 litres were needed to pay one man's weekly wage in 1991–92 compared to 802 litres in 1979–80, an increase of 57%.

The trend towards greater specialisation in dairy farming is illustrated by Table 2.1. During the period 1955 to 1992 the number of producers fell from 142,792 to 29,439 and at the same time size of the average herd increased from 17 to 74 cows.

DIVERSIFICATION

Despite the trend towards specialisation there are still good reasons why many dairy farms do not concentrate solely on milk production.

Table 2.1 The changes in the number of registered producers and cows 1955–92 in England and Wales

Year	*No. producers*	*No. of dairy cows (thousands)*	*No. cows per producer*	*Yield per cow (litres)*
1955	142,792	2,415	17	3,065
1960	123,137	2,595	21	3,320
1965	100,449	2,650	26	3,545
1970	80,265	2,714	34	3,755
1975	60,279	2,701	45	4,070
1980	43,358	2,672	61	4,715
1985	37,815	2,580	68	4,765
1990	31,510	2,324	73	5,070
1992	29,439	2,180	74	5,175

MMB Dairy Facts and Figures.

1. Lack of adequate buildings and/or lack of land suitable for dairy cows are major reasons. The former may only be a temporary restraint and can be overcome given adequate capital.
2. Problems encountered in disposing of slurry restrict the number of cows that can be kept and may lead to the introduction of an enterprise such as spring cereals.
3. Many farms remain diversified simply because the farmer likes it that way even if this may mean a lower overall profit at the end of the day. This is particularly true when it comes to rearing dairy replacements.
4. There is a case for diversification to reduce risks and to even out profits from year to year. The risk of introducing disease is regarded by many farmers as a good reason for rearing youngstock. Feed prices vary considerably from year to year relative to milk price and this leads to considerable fluctuation in profits. Growing cereals is a way of reducing the effect of these price changes and consequent variations in profits from year to year.
5. The success or otherwise of many businesses, not only in farming, depends on the efficient disposal of by-products. The calf is a major by-product of the dairy and its effective disposal may necessitate or be an added reason why a beef enterprise should be introduced.
6. Surplus labour at certain times of the year or all the year may lead to the introduction of a new enterprise such as calf rearing or a small pig enterprise so as to justify maintaining this labour on the farm.

7. The ability to grow highly profitable crops such as potatoes and winter wheat will also lead to diversification. In certain parts of the country the introduction of such a crop is particularly advantageous as it allows double cropping of the land, for example early potatoes followed by kale or stubble turnips.
8. The introduction of milk quotas has given an added impetus to dairy farmers to diversify their business due to the expense involved in acquiring additional milk quota, either to lease or to buy.

COMPETITIVE, SUPPLEMENTARY AND COMPLEMENTARY ENTERPRISES

The discussion to date has centred round the reasons for specialisation and diversification. In practice, it is up to the individual farmer to decide on the right balance of enterprises at his disposal. In this connection it is usual to designate enterprises as being competitive, supplementary or complementary to each other in relation to their demands for land, labour and capital.

A calf-rearing enterprise on a dairy farm is competitive with the dairy cows in its need for capital, but is not competitive with land. It may be feasible to carry this enterprise without any additional labour and utilise buildings which would otherwise remain idle; ie, it is a supplementary enterprise. On many dairy farms there is an area of land not suited to grazing by dairy cows and this can be grazed by dairy heifers without competition to the dairy herd. Whether or not this competition exists between enterprises, whether it be for labour or capital, needs to be assessed most carefully when deciding on the cropping and stocking of the farm.

An enterprise is said to be complementary when it makes a contribution to the success of another. The rearing of dairy heifers contributes to the success of a dairy herd if the heifers reared perform better than purchased replacements. Some of the merits of ley farming are questioned but most people would expect crops such as potatoes or winter wheat to yield better than normal following a ley and/or to be grown at less cost due to savings in fertilisers. In this case the dairy herd can be said to have a complementary effect on the cereal crops.

One of the arts of management is to be able to devise a farming system that exploits the complementary and supplementary relationships between enterprises and minimises direct competition for resources. This applies whether one is considering the role of cereals on a dairy farm or the place of dairy cows on an arable farm. The term 'art of management' is used because it is very difficult to quantify these relationships and express them in monetary terms.

THE PRINCIPLE OF DIMINISHING RETURNS

This principle has special application when we are considering the yield level at which to aim, whether it be milk yield per cow or grass yield per hectare.

We can expect to increase the yield per cow by providing greater and greater quantities of high energy or concentrate feed. Similarly the yield of grass can be increased by using more and more nitrogen. However, a point is reached when additional inputs tend to produce a diminishing increment of output for each additional input. Eventually a point is reached where additional input leads to loss of output. This relationship is illustrated in Figure 2.1.

Figure 2.1 The principle of diminishing marginal returns

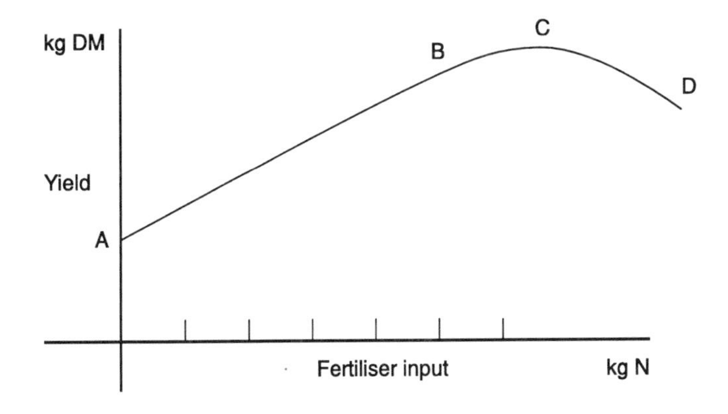

Between A and B there is a linear relationship, ie the increase in yield for a given additional input is constant. Between points B and C there is a diminishing return for additional inputs and yield eventually falls between points C and D.

The maximum yield occurs at point C but the optimum yield occurs at a point between B and C. At this point the value of the additional or marginal increment of output is equal to the value of the additional or marginal increment of input. It is also the point at which the margin of yield over inputs is at its maximum.

It is relatively easy to state this principle but in practice it is difficult to evaluate when the optimum point occurs due to the complex technology of the dairy farm enterprise. Hence the conflicting arguments regarding the optimum level of concentrate feeding to dairy cows. One school of thought will put the optimum at less than one tonne of feed per cow per annum with yields in the region of 5,000 litres per cow, representing a concentrate use of less than 0.20 kg per litre, whereas another will put the optimum level of feeding at 0.35 kg per litre with expected yields in excess of 6,500 litres.

A similar case exists in relation to nitrogen inputs on grassland. It is generally accepted that a linear relationship exists between yield and nitrogen input up to about 300 kg per hectare. Considerable conflict of opinion, however, occurs when it comes to the effect of frequency of cutting on yield. Fewer cuts will increase the total yield of dry matter but reduce the quality of grass conserved. In this instance it is very difficult to evaluate the monetary value of the marginal output of, say, silage for the additional marginal input of labour and machinery.

This need for additional inputs of labour and machinery may also occur at the same time as they are needed for another enterprise, for example third-cut silage often clashes with the cereal harvest. The problem to be answered in this case becomes the wider one of: 'Should we use our labour and machinery to harvest silage or to harvest cereals?'

THE PRINCIPLE OF EQUI-MARGINAL RETURNS AND OPPORTUNITY COSTS

To answer the question posed above we need to work out the value of the third-cut silage less the cost of making it, and compare this to the cost of having the cereals combined by a contractor and the adverse effect this may have on the value of the cereal harvest. As a result we would hope to arrive at a decision that maximised the profit from the farm as a whole, but in doing so we would have to accept a lower net return from one enterprise. This lower return from the one enterprise would represent the opportunity cost of making a profit from the other. Some economists would argue that opportunity costs are the only real costs. For example, the real cost of growing a silage crop on an arable farm is the margin that could be obtained from growing a cereal crop.

This principle of opportunity costs is fundamental to the approach we should make in our decision-making in farm management. Partial budgets and gross margin budgets represent methods of application of this principle and are discussed in more detail in Chapter 8. They are the means whereby we hope to achieve the optimum balance of enterprises and methods of production on the farm. This situation will obtain when the marginal returns from all enterprises are equal.

PRINCIPLES OF SUBSTITUTION

When considering the principle of diminishing returns we simply considered the effect at the margin of using more and more of an input such as nitrogen or concentrates in relation to the yield from a given area of land or a dairy cow. However we also need to consider whether one product can be substituted for another to produce a given yield, the objective being to substitute a low-cost input such as grazed grass for concentrates.

In the feeding of a dairy cow we are not simply considering the effect of changes in the amount of concentrates fed, we are also changing the amount of other feeds fed and substituting silage or other bulk feeds for concentrates and vice versa. This principle is often forgotten by so-called experts who draw erroneous conclusions from figures showing the relationship between milk yields and concentrate inputs that assume 'other things are equal', whereas in fact they are quite different.

To date, much of the discussion has centred round inputs such as feed and fertilisers but the substitution of capital in terms of machinery equipment and buildings for labour is of equal, if not greater importance. Again, however, this cannot be considered in isolation from the rest of the farm. It may be profitable to spend capital on machinery and equipment to speed up the rate of silage-making but even more profitable to spend it on more cows and/or improving the milking facilities. This takes us back once again to the principle of equi-marginal returns and opportunity costs, which should be central to our decision-making process in the preparation of our plans and policies for the farm.

If we understand and use this principle effectively we will achieve a good balance of enterprises and a balanced method of producing the main product, ie milk from our farm. Whatever the problem and whatever the challenge we need to ask questions such as 'Is there a better, less costly way of doing this job?' 'Do I need to do the job now or would it be better to do something else?'

However, we must not spend all our time thinking what to do or how to do it; that is the privilege of academics! We have to make a decision, get on and do the job or tell someone else to do it. Later, however, we must not forget to check the job was done in the way intended and yielded the result expected. This will make the decision much easier next time.

CHAPTER 3

Farm Accounts as an Aid to Management

HISTORICAL BACKGROUND

Prior to the 1939–45 war farms did not need to prepare accounts for tax purposes as income tax was charged on a per hectare basis. During the war this system came to an end and farmers became obliged to prepare tax accounts to be drawn up and certified as being correct by a fully qualified accountant.

This resulted in the development of accounting methods which were designed only to supply information for taxation purposes and these were of little or no use for management. The need to keep these accounts for tax purposes also engendered a deep suspicion and dislike of accounting in the farming community.

Farm accounts and record-keeping was taught in colleges for many years as a subject which was an end in itself and as such it was unpopular with staff and students alike. Before the introduction of farm business management techniques in the late 1950s and early 1960s it was not realised that accounts could and should provide the material for a detailed study of the dairy farming business.

The growing awareness of the value of accounts as an aid to management was largely the result of the work of the farm management specialists appointed to the staff of the agricultural economics departments of universities responsible for collecting financial information from farms for price review purposes. Each year these universities produced financial data which was used to establish the average net farm income of British farmers. This formed an essential part of the Annual Price Review Negotiations for many years and it is still the main source of financial data for the Government in its EU negotiations.

These farm management specialists included John Nix of Wye College. He recognised the need for a Farm Management Pocketbook to provide financial data for farmers and advisers. This it has done to good effect over many years and is now in its 24th edition. Data from this handbook is quoted in Chapter 5.

The data provided by the agricultural economics departments is fully authenticated and provides an undisputed source of reliable information on the economics of dairy farm production and other farm enterprises. The information is derived from accounts collected from bona fide farms all over the country.

In the late 1960s the Ministry of Agriculture introduced a Farm Business Recording Scheme. Under this scheme grants were paid to farmers for the production of management accounts providing these were presented in a standardised form. The Grant Aid Scheme did much to stimulate the use of accounts as an aid to management. Colleges set up courses to train farmers and farm secretaries to produce accounts suitable for management purposes and this work has continued to expand and develop over recent years. Commercial organisations have set up management costing schemes and foremost amongst these, from a dairy farmer's point of view, was the work of the Milk Marketing Board's Farm Management Services Department, now part of Genus.

STANDARDISATION OF ACCOUNTING METHODS AND MANAGEMENT TERMINOLOGY

Over the years, it is fair to say that advisers and farmers have come to realise that farmers need two sets of accounts, one for tax purposes and one for management purposes. This need arises from the need to compare 'like with like' when comparing an individual farmer's results to those achieved by other farmers. It is important to use the same terminology if confusion is not to arise when communicating with other farmers and advisers. For these reasons the Ministry of Agriculture has produced a *Glossary of Terms and Definitions Used in Farm Management*, and this is reproduced in Appendix 1.

Before discussing these various terms and definitions we need to consider the format of a farm trading, or profit and loss, account. A simple example is shown in Table 3.1. This account shows a profit of £30,000 after charging £6,000 depreciation and including £1,000 notional income. The latter represents the value to the farmer of the produce his family has consumed during the year, such as milk, and

Table 3.1 Example of farm trading or profit and loss account

	£		£
Opening valuation	95,000	Closing valuation	100,000
Trading expenditure	130,000	Trading revenue*	160,000
Depreciation	6,000	Notional income	1,000
Profit	30,000		
	261,000		261,000

* Or income.

also takes into account the rental value of the farmhouse and the value of other private consumption items paid by the business during the year. In the example the trading revenue exceeds the trading expenditure by £30,000, that is £160,000 less £130,000. This surplus of trading revenue over expenditure is not the same as the trading cash surplus as account has to be taken of the change in creditors and debtors. How revenue and expenditure relate to cash flow is illustrated in Table 3.2.

Table 3.2 Trading revenue and expenditure

	£		£
Trading payments	135,000	Trading receipts	159,000
Add closing creditors	10,000	Add closing debtors	16,000
	145,000		175,000
Subtract opening creditors	15,000	Subtract opening debtors	15,000
TRADING EXPENDITURE	130,000	TRADING REVENUE	160,000

The surplus of trading revenue over expenditure is £30,000 but the trading cash surplus, that is trading receipts less trading payments, is only £24,000. The cash surplus is £6,000 less than the revenue/expenditure surplus because creditors have been reduced by £5,000 and debtors have increased by £1,000. It is important to distinguish and to remember the difference between receipts and payments on the one hand and revenue and expenditure on the other, especially when talking to bankers.

When preparing farm accounts it is most important to appreciate the enormous effect that subjective judgements regarding the valuations of livestock, and other commodities on hand, at the beginning and end of the year can and do have on the profit. The same relative values for a commodity should be used at both the beginning and end of the year unless there has been a real change in values, so as to avoid showing profits (or losses) that are simply due to differences in valuation judgements. On dairy farms it is also important to avoid showing wide fluctuations in profits (or losses) that are simply due to changes in the value of home-grown forage supplies on hand between the beginning and end of the year.

In the trading account example used earlier (Table 3.1) the valuation has increased by £5,000, that is from £95,000 to £100,000. This could simply be due to an increase of £50 per head in the value of 100 dairy cows, the numbers on hand being the same at the beginning and end of the year, or it could represent an extra 10 cows at £400 per head plus an extra 5 heifers at £200 per head.

When analysing a farmer's financial results it is necessary to look thoroughly at the valuation details to see if they are masking the true result. Unfortunately this is often not feasible, particularly if the only information available is the income tax accounts, as the valuation produced by the valuer may have been arranged to produce a given result rather than vice-versa.

Valuations for tax purposes are usually produced on the basis of cost of production or 'deemed cost' (see Chapter 14 for definition). Valuations for management purposes need to be produced, based on market values if they are to show the true result. Valuations on this basis are not desirable from a taxation point of view so in most instances it is necessary to have two valuations, one for tax purposes and one for management purposes.

Finally, before leaving valuations a few comments need to be made about dairy cow values and the impact inflation has had on these values. At the present time (1994) a newly calved heifer entering the herd has a value in the region of £1,200 whereas the average cull cow (including casualties) has an average value in the region of £500. A figure midway between these two, ie in the region of £850 per head, is therefore an appropriate value for a herd at the present time. Ten years ago the figure calculated on the same basis would have been in the region of £400. Just one year ago the figure would have been in the region of £550.

Dairy cow values have tended to keep up with inflation over the years and it is important to show the effect of this 'hedge against inflation' in the management accounts. If this inflation hedge is not included, the relative profitability of dairy farming compared to, say, arable farming is underestimated. When the final management accounts are prepared this 'appreciation' in unit stock values should be shown as a separate item, ie breeding livestock appreciation.

Having discussed the problems of livestock values and inflation we need to look briefly at the problems inflation has brought in relation to the calculation of machinery and equipment depreciation. Traditionally this has been calculated on an historic cost basis, usually on a reducing balance method. For example:

	£
Tractor purchase in 1990	16,000
Depreciation 1st year @ 25%	4,000
Written-down value in 1991	12,000
Depreciation 2nd year @ 25%	3,000
Written-down value in 1992	9,000
Depreciation 3rd year @ 25%	2,250
Written-down value in 1993	6,750
Accumulated depreciation	9,250
Initial purchase price	16,000

Depreciation is meant to cover the cost of replacing the tractor when this becomes necessary. The accumulated historic depreciation of the tractor in the above example is £9,250 and its written-down value is £6,750. The purchase price of an equivalent replacement in 1994 is likely to be in the region of £20,000 and the trade-in value of the old tractor is likely to be in the region of £8,000. The old tractor is undervalued in the books by £1,250 and 'tax' would be payable on this paper profit if a decision was taken to give up farming and sell the tractor. The accumulated depreciation, on the other hand, of £9,250 is £2,750 less than the £12,000 (£20,000 minus £8,000) required to replace the tractor if a decision is taken to continue in farming.

There is no simple solution to this problem. Cambridge University, for example, now produce their reports with depreciation calculated by both historic cost and current cost methods. Whichever method is used, the result shown is very subjective and open to error.

The only really worthwhile advice that can be given is to treat all depreciation figures with caution and to accept that at best the depreciation figure is only an estimate. In particular, it should be noted that depreciation figures produced on the historic cost basis will grossly underestimate the capital sum required to meet current machinery replacement requirements.

What matters in practice is the amount being spent/planning to be spent on capital items. In management accounts and budgets, therefore, it is often a very good idea to produce a profit before depreciation and then deduct from this the actual cash that is planned to be spent, or has been spent in the previous year.

PROFIT DEFINITION

Having discussed some of the principles involved in the preparation of a farm trading or profit and loss account we now turn to the adjustments that need to be made to the accounts so that 'profit' can be accurately defined and standardised so that the results can be compared to those achieved by other farms.

The basic objective in making these adjustments is to arrive at the net farm income which is defined as the 'return to the farmer and his wife for their manual labour, management and tenants' capital invested in the business'. Any interest charges they pay are deducted; if they are owner-occupiers a rented value is included for the land owned and ownership expenses incurred in respect of this land are excluded. Net farm income is defined as the return to the 'farmer and his wife' only, so to arrive at this figure the value of work done by unpaid family labour has to be added to the actual paid wages. Finally, any non-farm expenses are extracted from the accounts and an addition is made for the value of notional benefits the farmer and his wife receive, such as the rented value of the farmhouse, the value of produce consumed, the value of electricity and fuel consumed in the house and costs saved by having the personal use of the farm car. These non-farm costs and notional benefits can be substantial and, on the smaller farm in particular, they often represent a large proportion of the total net farm income.

At best these estimates are only approximations. Consequently it is often difficult to ascertain the true profit made by a small farm and the true costs, particularly power and machinery costs, involved in running a farm are difficult to gauge due to the effect on these of private cars and fuel consumption.

Having calculated net farm income it is usual to subtract the value of the farmer's and wife's manual labour in order to arrive at the management and investment income which represents the reward to management and return on tenant's capital invested in the farm. It is unfortunate that this is the final approved definition of profit as this still measures the return before making a charge for management. Many farm businesses are now run by salaried managers and it would be helpful to have standard 'investment income' data.

The salutary effect the changes outlined above can have on the 'profit' made by the farm is illustrated in Table 3.3. The farmer concerned is an owner-occupier on 60 hectares. He is reasonably complacent because he and his son, who is a partner in the business, have made a 'profit' according to his income tax account of £24,000. However after making the adjustments referred to above, we find that this is reduced to a management and investment income of –£2,000.

Table 3.3 Arriving at management and investment income

	Total (£)	*£ per hectare*
Profit according to income tax accounts	24,000	400
ADD:		
Interest charges paid to bank	1,800	30
Ownership expenses	600	10
Notional income	1,600	27
	28,000	467
SUBTRACT:		
Rental value	9,000	150
Value of son's unpaid manual labour (including 14% national insurance)	10,500	175
Value of farmer's unpaid manual labour	10,500	175
	30,000	500
MANAGEMENT AND INVESTMENT INCOME	–2,000	–33

ACCOUNT ANALYSIS

So far our discussion has concerned the various adjustments we need to make to the farm accounts and the care that has to be taken in their preparation if one is to be able to compare the profit made by one farm to another and draw sensible conclusions as to their relative profitability.

Universities and other organisations produce standard performance data and the results shown in Table 3.4* are for a group of dairy farms costed by Manchester University. If we look at this data we see that it is not presented in the form of a conventional trading account as described earlier in this chapter but in terms of gross output, variable inputs (costs), gross margin and fixed inputs (costs) per hectare. If we look at the results more carefully we will also notice that the profit (or management and investment income) equals the gross margin less the fixed inputs and that the gross margin equals the gross output less the variable inputs.

To explain these terms it is now proposed to take a relatively simple example trading account for a 40 hectare dairy farm and show how this is modified to produce gross outputs, fixed costs, variable costs and gross margins. Having read this section the reader is then recommended to study the glossary of terms in Appendix 1. The trading result for our example farm is summarised in Table 3.5. To arrive at the variable costs and fixed costs the expenditure has to be adjusted where necessary for the changes in the valuation, as illustrated in Table 3.6.

The next step in the analysis is to calculate the gross margins achieved from the individual enterprises, in this case the dairy cows and the dairy replacements. To do this, one needs to know the number and value of the dairy heifers transferred into the dairy herd and the number and value of the calves transferred from the dairy herd to the dairy replacements enterprise. In addition one has also to be able to allocate the variable costs, ie feedingstuffs, livestock sundries, etc, between these two enterprises.

Finally, one also needs to show the average number of stock carried throughout the year so that the results can be expressed on a per-head basis. Given this information the gross margin results for this example farm can be analysed as shown in Table 3.7

The allocation of the feed and livestock sundries to the dairy cows on the one hand and the replacements on the other is relatively simple providing the necessary records are kept during the year. The allocation of seeds, fertiliser and other crop variable costs is much more difficult. This is usually resolved by making an allocation on a livestock unit basis as illustrated in Table 3.8. There are inherent weaknesses in this method of allocation and this is discussed later.

Finally, a summary of the accounts can be produced, as shown in Table 3.9.

* Table 3.4 and the remaining tables referred to in this chapter are shown on pages 35–43.

PROVIDING THE INFORMATION

To produce accounts in the form described, a good bookkeeping system is essential but the time spent on it need not be substantial if properly organised. To produce accounts in the form described for a farm of 40 to 100 hectares, a well-trained secretary should only need to make 12 half-day monthly visits to the farm.

A cash analysis book is essential and the headings used in this book should conform to those shown in the trading account earlier in this chapter. A farmer setting up such a system for the first time would be advised to employ a secretary and/or to purchase the *NFU/ADAS Record Book* which has been designed to record the information required for farm business analysis.

To provide the information required for gross margin analysis there is a need for physical as well as financial information and records have to be kept so that variable cost items can be allocated to individual enterprises.

The biggest problem is usually to do with the allocation of concentrate feeds. If there is only a dairy herd and a replacement enterprise and all food is purchased there is little or no difficulty. The problem increases when there are other livestock enterprises, home-grown grain is fed and rations are made up on the farm. In this case some form of feed recording system is essential. The system needs to be designed to meet the needs of the individual farm and must be designed so that the records of what is said to have been fed can be reconciled to actual deliveries and stocks on hand at the beginning and end of the month. This feed record should also be used to compare what has actually been fed to any group of stock to what was intended to be fed.

USE OF COMPUTERS

Computers are a very topical subject and the reader may well expect a long dissertation on the use of computers in this book. This, however, will not be the case as the use of computers needs to be kept in perspective.

The time required to write up the books for a 40 to 100 hectare farm on a manual system is only in the region of 6 days or 48 hours per annum. The installation of a computer is not going to bring about significant savings on such a farm from a costings point of view. It is increasingly likely, however, that the secretary or others involved in the farm costings will make use of a microcomputer in his or her work. The dairy farmer can also expect to make increasing use of computerised services in the day-to-day control of the feeding and breeding of the dairy herd.

The larger farm finds it easier to justify the outright purchase of a computer at the present time. Even so, the farmer has to consider carefully whether or not the capital would be better spent, for example on the installation of automatic cluster removers in the parlour, rather than on the purchase of a computer system.

The would-be purchaser also needs to be aware of the dangers of obsolescence. The speed of change in computer technology is very rapid and products on the market today and companies in this market could well have disappeared or have been superseded by better products in two or three years' time. The general advice to most small businesses is to wait and see what develops before purchasing a computer. In the meantime consider making use of a bureau service to gain experience in the kind of service a computer can provide.

COMPARATIVE ANALYSIS

Having analysed and processed our accounts into gross margins and fixed costs, we are now in a position to compare our results to those achieved by other farms costed on the same basis. We can also compare the results to those we hoped to achieve and this forms the basis of the budgetary control system described later.

When comparing our results to standards we need to be aware of the limitations of the standard data and the need to try to ensure that like is being compared to like. Standard data for the whole farm is normally presented on a per-hectare basis. The gross margins, fixed costs and profitability levels per hectare tend to be higher the smaller the size of the farm. It is important therefore to try to compare the results for a particular farm to those of other farms of a similar size.

In periods of inflation and rapidly changing levels of profitability it is also important to compare results to those achieved by other farms in the *same year*. This can pose a problem as the comparative data is not usually available until some 9 to 15 months after the end of an accounting year.

It must also be borne in mind that the standard data covers a very wide range of performance. The results in most surveys are therefore divided to show the results for the top 25–30% of farms as well as the results for the average (see, for example, the Manchester University data quoted in this chapter in Table 3.4).

Despite these limitations, much can be gained from a comparison of the results from an individual farm to standards. Much can also be learned by studying the differences between the most profitable and average farms within a group. Note, for example, that the most profitable farms in the Manchester University Survey achieve much higher gross margins but incur little more expenditure on fixed costs than the average farm.

When comparing the results for an individual farm to standards, the first step is to look at the results for the farm as a whole to see whether the individual farm has a higher or lower net farm income or management investment income than that achieved by the comparative farms. The next step is to continue the comparison to see whether this advantage or disadvantage is due to differences in the fixed costs structure or to differences in the gross margin. One of the major objectives in the comparison is to determine where scope lies to increase profits. If fixed costs are higher than average this will suggest that these should be examined to see if they can be reduced. On most farms, however, the main scope for improving profits is by increasing the gross margin and this aspect is discussed in more detail in the next chapter.

FIXED AND VARIABLE COSTS

Before moving on to a detailed discussion on profitability factors, an explanation is necessary of the concept of fixed and variable costs (for formal definitions see Appendix 1).

The items usually treated as fixed costs were detailed earlier in this chapter. Fixed costs are those costs which do not change as the result of a small change in the organisation of a business. They are also difficult to allocate to an individual enterprise from a costings point of view. If, for example, a farmer decides to keep 75 cows instead of 65, it is unlikely that this would have any direct effect on the cost of labour, rent, rates, power and machinery costs or on general overheads. It would, however, immediately result in a need for more concentrate feeds, more quota and an increase in sundry costs such as artificial insemination fees. Whether it would immediately lead to an increase in seed and fertiliser cost is debatable, but in the long term the cost of these items would tend to rise in proportion to the increase in cow numbers.

The significance of the above is that the effect on the profit of an additional 10 cows can be assessed without needing to know what the fixed costs are per cow. The increase in the profit due to keeping the additional cows will be equal to the gross margin that can be expected from these 10 cows, net of the additional cost of leasing quota. Keeping an extra 10 cows will lead to a reduction in the other fixed costs per cow but the total fixed costs will remain the same.

There is one exception to this rule, however, and that is the effect of the increase in cow numbers on finance charges. Finance charges are usually treated as part of fixed costs for convenience, as they cannot be readily allocated to an individual enterprise, but in a borrowed capital situation a small increase in cow numbers will immediately lead to an increase in finance charges. The increase in profit from the additional cows is therefore equal to the gross margin they produce less the interest charges on the capital borrowed to purchase the cows.

It is important to remember that the effect on the profit of a change in cow numbers is not so easily assessed if the change in numbers is more substantial. An increase in cow numbers from 65 to say 95 would probably also necessitate changes in building and equipment and may lead to the need for more overtime. Certain overhead costs, however, would stay the same and the increase in fixed costs would not be proportionate to the increase in cow numbers.

This concept of fixed and variable costs is fundamental to the understanding of farm business management principles and budgeting methods and is discussed again later in the book. It also explains why very few complete costings are carried out in farm business management. Knowing how much 'profit' is made from an individual dairy cow is of little value as the change in profit of the whole farm as a result of changing dairy cow numbers is not proportionate to the profit made per cow. This is the basic reason for the development of the gross margin costing system which has the added advantage of being much less time-consuming than a full enterprise costing system.

GROSS MARGIN LIMITATIONS

The gross margin concept is very useful but, as mentioned earlier, there are difficulties on a dairy farm in allocating the forage costs between the dairy cows on the one hand and the youngstock on the other. There are also problems with the allocation of the value of the calves between the dairy cows and the youngstock and also a problem regarding the value that should be used for down-calved heifers entering the herd.

Experience over the years has shown that the gross margin

attributed to the youngstock on the one hand compared to the cows on the other, is very dependent upon subjective judgement and one begins to wonder whether or not this division is worth the time and effort involved.

Increasingly, therefore, the tendency is to look at the gross margin from the dairy herd as a whole, looking firstly at the margin over concentrates (MoC), and then secondly at the cattle output that is achieved by the dairy herd and youngstock enterprise combined.

It is still, however, very useful to have rule-of-thumb standards, eg, that the gross margin that can be expected from a cow is £900 and the gross margin that can be expected from a youngstock animal is £200. This is very helpful when building up rule-of thumb budgets for a farm.

The introduction of milk quotas has meant that the gross margin from a cow depends upon whether it is producing the quota allocated to the farm or whether it is producing a gross margin after the leasing costs have been deducted. Here the rule-of-thumb budget gross margin is still useful, eg £540 per cow, £900 less £360 re leasing costs.

THE FARM BALANCE SHEET

The farm balance sheet performs two functions. Firstly, as its name implies, it provides a means whereby the books or accounts for a particular year can be balanced. The objective in this instance is to reconcile the profit shown in the trading account with the overdraft at the bank and with the change that has taken place in other assets and liabilities. For the farm described earlier in this chapter, the balance sheet statement could appear as shown in Table 3.10. This statement shows an increase in the net capital or net worth of £20,000 as the private drawings including tax payments are £20,000 less than the profit made during the year.

The second function of the balance sheet is to show in some detail the assets and liabilities of the business and how the net worth is determined.

When presenting a balance sheet it is advantageous to have the assets listed in descending order of liquidity, that is the ease with which the capital invested in them can be realised. The liabilities should also be listed in ascending order in which they are due for payment, that is with trade creditors and the bank overdraft at the top and long-term loans at the bottom. It is also advantageous to show the balance sheets for several years alongside each other so that the trend in net worth and in the ratio of assets to liabilities can be readily identified. A form that can be used for this purpose is shown as Table 3.11.

The figures shown in Table 3.11 for 1988 and 1989 are for a 80 hectare tenanted farm. Between 1988 and 1989 there was an increase in total assets from £98,000 to £110,000. Short-term liabilities increased by £5,000 and there were no long-term liabilities so the net worth was up by £7,000. The business was in a strong liquidity position as current assets were well in excess of short-term liabilities. There was, for example, £8,000 saleable crop produce on hand at 31 March 1989 and the farmer could use this if he wished to repay most of the bank overdraft.

1989–90 is assumed to be another profitable year. The net worth went up by a further £8,000 and the farmer purchased 20 hectares of adjoining land for £100,000. £5,000 was funded from profits, £80,000 by an AMC loan and £15,000 by an increase in the overdraft. At this stage the liquidity position still seemed reasonable as the current assets still exceeded the short-term loans. By 1991, however, a different picture had emerged. Expenditure had been incurred on buildings and equipment and on breeding livestock and the total fixed assets had gone up by a further £19,000. An attempt to fund some of this out of profits had been unsuccessful and the liquidity position had deteriorated. There was no saleable crop produce on hand. Consumable stores were down as funds have not been available to purchase fertilisers in advance, as was the case in previous years. Current assets were £9,000 less than short-term loan liabilities. This business has cash flow problems as too great a proportion of the investment was in fixed assets. The position could be eased by replacing some of the short-term loans by long-term loans but the basic problem was that the service charges, that is the interest charges on the borrowed capital, were more than the business could afford. To avoid getting into this situation careful budgeting and planning is essential.

Table 3.4 Financial performance (£ per hectare), lowland dairy farms, over 50 hectares

	1990/91	*1991/92*	
	Average	*Average*	*High profit*
	£/ha	*£/ha*	*£/ha*
Gross output per hectare			
Arable crops (current year)	61	64	91
Forage crops and by-products	6	37	31
Dairy herd	1,695	1,778	2,174
Other cattle	178	170	144
Sheep and wool	25	21	
Miscellaneous	44	40	27
Total gross output (A)	2,009	2,110	2,467
Variable inputs per hectare			
Feed:			
Purchased concentrates	409	389	418
Home-grown concentrates and milk	24	25	43
Purchased bulk feeds and keep	58	63	40
Vet and medicines	44	47	50
Other livestock costs	69	77	77
Seed	12	13	18
Fertilisers	112	110	126
Sprays	10	16	24
Other crop costs	25	23	38
Total variable inputs (B)	763	763	834
Gross margin (C) (=A–B)	1,246	1,347	1,633
Fixed inputs per hectare			
Machinery	334	367	426
Labour incl. farmer and spouse			
paid	163	173	230
unpaid	168	191	190
Land expenses:			
Repairs and maintenance	31	39	32
Rent and rates	155	165	161
General overhead costs	96	109	108
Total fixed inputs (D)	947	1,044	1,147
Management and invest. income (E) (= C–D)	299	303	486
Farmer and spouse labour (F)	145	158	153
Net farm income (E+F)	444	461	639

All figures are net of breeding livestock appreciation.
Crop output: Arable output and adjustment includes farm consumed.
Forage and by-product output comprises sales and valuation changes only.

Table 3.5 Example trading account

Opening valuation:	£	*Closing valuation:*	£
Dairy cows: 60 @ £500	30,000	Dairy cows: 64 @ £500	32,000
		Breeding, livestock appreciation: 64 @ £200	12,800
Dairy replacements: 40 @ £300	12,000	Dairy replacements 42 @ £400	16,800
Hay and silage	2,000	Hay and silage	2,200
Seeds	200	Seeds	Nil
Fertilisers	1,000	Fertilisers	6,000
Purchased feed	4,000	Purchased feed	3,600
Fuel and oil	200	Fuel and oil	100
TOTAL	49,400	TOTAL	73,500
Expenditure	£	*Revenue*	£
Dairy cow	1,000	Milk	73,800
Feedingstuffs	18,600	Cull cows	6,200
Vet and medicines	2,200	Calves	5,000
AI and dairy expenses	2,600	Youngstock	1,200
Other livestock expenses	1,200		
Seeds	200		
Fertilisers and lime	10,000		
Other crop expenses	1,500		
Wages and National Ins.	8,000		
Fuel and oil	1,900		
Electricity	2,500		
Machinery repairs/tax/insurance	4,500		
Contract	2,000		
Rent	3,200		
Property repairs	1,000		
General insurance	1,000		
Professional fees/office expenses	1,000		
Sundry overheads	600		
AMC loan interest	3,000		
Bank charges and interest	2,400		
TOTAL EXPENDITURE	68,400	TOTAL REVENUE	86,200
Depreciation:		*Notional income:*	
Machinery and equipment	8,000	Milk	400
Tenant's improvements	800	Rental value of house	500
		Use of car	1,000
	8,800		1,900
Profit (inc. BLA)	35,000		
	161,600		161,600

Table 3.6 (i) Variable costs

	£	£
(a) Feedingstuffs: Expenditure	18,600	
PLUS opening valuation	4,000	
LESS closing valuation	(3,600)	19,000
(b) Livestock sundries: Vet and Medicines	2,200	
PLUS AI and dairy expenses	2,600	
PLUS other livestock expenses	1,200	6,000
(c) Seeds: Expenditure	200	
PLUS opening valuation	200	
LESS closing valuation	Nil	400
(d) Fertiliser and lime expenditure	10,000	
PLUS opening valuation	1,000	
LESS closing valuation	(6,000)	5,000
(e) Other crop expenses	1,500	
PLUS opening valuation	Nil	
LESS closing valuation	Nil	1,500
TOTAL VARIABLE COSTS		31,900

Table 3.6 (ii) Fixed costs

	£	£
(a) Labour:		
Wages and National Insurance	8,000	8,000
(b) Power and machinery:		
Fuel and oil expenditure	1,900	
PLUS Opening valuation	200	
LESS Closing valuation	(100)	
	2,000	
Electricity	2,500	
Machinery repairs/tax/insurance	4,500	
Contract	2,000	
Machinery and equipment depreciation	8,000	19,000
(c) Property charges:		
Rent	3,200	
Property repairs	1,000	
Tenant's improvements depreciation	800	5,000
(d) Sundry overheads		
General insurance	1,000	
Professional fees/office expenses	1,000	
Sundry overheads	600	2,600
(e) Finance charges		
AMC loan interest	3,000	
Overdraft interest and bank charges	2,400	5,400
TOTAL FIXED COSTS		40,000

NOTE: To be precise, one would deduct the rental value of the house and the use of the car from the fixed costs. However, as these estimates are so subjective, it is considered more practical to show these as a separate item on the summary of gross margins. See Table 3.9.

Table 3.7 Calculating individual enterprise gross margins

Average no.	*Total* 104	*Dairy cows* 62		*Youngstock* 42	
Gross output	£	£	*No.*	£	*No.*
Closing valuation	48,800	32,000*	64	16,800	42
Cow sales	6,200	6,200	13		
Calf sales	5,000	5,000	35	—	
Youngstock sales	1,200			1,200	3
Transfers out	19,300	3,300	25	16,000	16
Sub-total A	80,500	46,500	137	34,000	61
Opening valuation	42,000	30,000	60	12,000	40
Purchases	1,000	1,000	1	—	
Transfers in	19,300	16,000	16	3,300	25
Sub-total B	62,300	47,000	77	15,300	65
			Per head		*Per head*
Cattle output (A–B)	18,200	–500	–8	18,700	445
Milk sales	73,800	73,800	1,190		
GROSS OUTPUT	92,000	73,300	1,182	18,700	445
Variable costs:					
Feedingstuffs	19,000	15,600	252	3,400	81
Livestock sundries	6,000	5,100	82	900	21
Seeds, fertilisers and crop sundries	6,900	5,020	81	1,880	45
Total variable costs	31,900	25,720	415	6,180	147
GROSS MARGIN	60,100	47,580	767	12,520	298
Margin over concentrates		58,200	938		
No. litres sold		369,000	5,954		
Milk price per litre		20.00p			
Feed cost per litre		4.22p			
Margin over feed per litre		15.78p			

* Excluding breeding livestock appreciation

Table 3.8 Allocation of crop variable costs

Livestock units	*No.*	*Livestock unit factor*	*No. units*
Heifers over 2 years	8	0.8	6.4
Heifers 1–2 years	16	0.6	9.6
Heifers under 1 year	18	0.4	7.2
			23.2
Dairy cows	62	1.0	62.0
TOTAL	—	—	85.2

Forage costs (40 hectare)	£
Seeds	400
Fertilisers	5,000
Sprays and sundries	1,500
TOTAL	6,900

Forage costs per livestock unit	£6,900 ÷ 85.2 = £80.98
No livestock units per hectare	85.2 ÷ 40 = 2.13

Allocation		£	*Hectares*	
To dairy cows	£80.98 × 62 =	5,020	29.1	(62 ÷ 2.13)
To dairy heifers	£80.98 × 23.2 =	1,880	10.9	(23.2 ÷ 2.13)
		6,900	40.0	

Table 3.9 Gross margin/fixed cost summary

	£	*Per hectare*	*Per litre*
GROSS MARGIN			
Margin over feed	58,200		15.77
Cattle output	18,200		4.93
	76,400	1,910	20.70
LESS Youngstock feed	3,400		
Livestock sundries	6,000		
Forage cuts	6,900		
	16,300	407	4.42
	60,100	1,503	
ADD Forage valuation change	200	5	
Notional income	1,900	47	
TOTAL GROSS MARGIN	62,200	1,555	16.85
FIXED COSTS			
Labour (paid only)	8,000	200	2.17
Power and machinery	19,000	475	5.15
Property charges	5,000	125	1.35
General overheads	2,600	65	0.70
Finance charges	5,400	135	1.46
	40,000	1,000	10.83
Profit excluding BLA	22,200	555	6.01
Breeding livestock appreciation	12,800	320	
Profit including BLA	35,000	875	

Table 3.10 Change in net capital (worth) over a year

	£
Net capital (worth) at start of year	80,000
ADD	
Profit for year	35,000
	115,000
DEDUCT	
Private drawings during year	12,000
Tax payments	3,000
	15,000
NET CAPITAL (WORTH) AT END OF YEAR	100,000

Table 3.11 Statement of assets and liabilities (four years—31/3/88 to 31/3/91)

	80 hectares: all tenanted		*80 hectares as tenant 20 hectares owned*	
	31.3.88	*31.3.89*	*31.3.90*	*31.3.91*
	£	£	£	£
Assets:				
Current				
Cash in hand and at bank	—	—	—	—
Debtors	8,000	8,000	8,000	10,000
Saleable crop produce	7,000	8,000	8,000	Nil
Consumable stores	6,000	6,000	6,000	4,000
Growing crops and tillages	5,000	6,000	6,000	8,000
Trading livestock	18,000	22,000	22,000	16,000
Total current assets	44,000	50,000	50,000	38,000
Fixed:				
Breeding livestock	36,000	40,000	45,000	50,000
Machinery and equipment	18,000	20,000	20,000	24,000
Buildings and fixtures	—	—	—	10,000
Land	—	—	100,000	100,000
Total fixed	54,000	60,000	165,000	184,000
TOTAL ASSETS	98,000	110,000	215,000	222,000
Liabilities:				
Short term loans				
Creditors	8,000	8,000	10,000	12,000
Bank overdraft	5,000	10,000	25,000	35,000
Hire purchase	—	—	—	—
Other short-term loans	—	—	—	—
Total short-term	13,000	18,000	35,000	47,000
Long-term loans				
Bank loans	—	—	—	—
AMC loans	—	—	80,000	80,000
Total liabilities	13,000	18,000	115,000	127,000
NET WORTH	85,000	92,000	100,000	95,000

CHAPTER 4

Whole Farm and Fixed Cost Profitability Factors

SYSTEM EFFICIENCY AND ENTERPRISE MANAGEMENT EFFICIENCY

The previous chapter has shown that the profit made by a farm equals the gross margin minus fixed costs. Chapter 1 described the functions involved in management, indicating that a farmer or a manager has basically two jobs to do: first, to decide on the farm policy and second, to put this into practice. It follows that the profit made on a farm depends on how well he does these two jobs. The best profits are made when there is a good farming system and this is effectively managed.

The relationship between farming profitability, gross margins, fixed costs and the concept of system efficiency and enterprise management efficiency is illustrated in Figure 4.1. This diagram shows how the various husbandry, economic management and other factors determine the profit of a dairy farm. The successful dairy farmer understands these rather complex interrelationships and is able to bring them together to make a success of the dairy farm business as a whole.

The effect on the farm profit of differences in system efficiency and enterprise management efficiency is illustrated in relation to a part-owned, part-tenanted 70 acre farm by the data set out in Table 4.1. Farmer A and Farmer B have the same farming system, in other words, the same number of cows and the same number of youngstock, but the profit made by Farmer B is only £4,060 compared to a profit made by Farmer A of £19,600. This very substantial reduction in the profit is the consequence of an assumed 10%

Figure 4.1 Factors determining farm profits

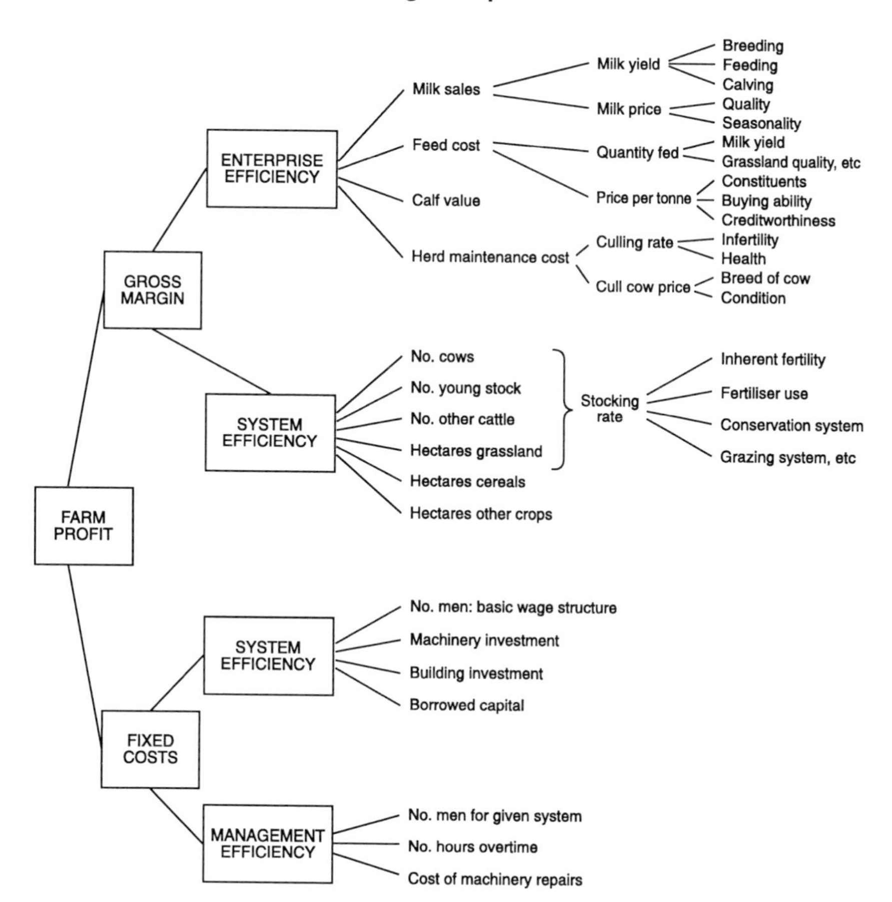

reduction in the gross margin and a 10% increase in the fixed costs, in turn due to an assumed lower level of management ability.

The lower level of management ability in relation to enterprise performance is fairly easy to comprehend, but that in relation to fixed costs is much more difficult. Experience has shown that some farmers are simply capable of containing fixed costs much better than others, and the only way that one can describe this is that they are 'good housekeepers'. In other words, they manage to avoid spending money unless it is really necessary.

Turning to the comparison between Farmer A and Farmer C: in this case it is assumed that they have the same level of enterprise efficiency and the same ability to control fixed costs. Farmer C, however, makes £9,600 less profit than Farmer A, simply because his farm carries 10% less stock. This is typical of the situation found

Table 4.1 The effects of different system and enterprise management efficiencies on the profitability of a 70 hectare dairy farm

	A Good system and good enterprise management			B Good system but poor enterprise management			C Poor system but good enterprise management		
	No.	£ per head	£	No.	£ per head	£	No.	£ per head	£
Gross margin:									
Dairy cows	100	800	80,000	100	720	72,000	90	800	72,000
Youngstock	76	200	15,200	76	180	13,680	68	200	13,600
		£ per hectare			£ per hectare			£ per hectare	
TOTAL GROSS MARGIN		1,360	95,200		1,244	85,680		1,223	85,600
Fixed costs:									
Labour inc. farmer		400	28,000		440	30,800		400	28,000
Power & machinery		360	25,200		396	27,720		360	25,200
Sundry overheads		100	7,000		110	7,700		100	7,000
		860	60,200		946	66,220		860	60,200
Profit before rent and finance charges		500	35,000		278	19,460		363	25,400
Rent		80	5,600		80	5,600		80	5,600
Finance charges		140	9,800		140	9,800		140	9,800
Rent equivalent		220	15,400		220	15,400		220	15,400
PROFIT MARGIN		280	19,600		58	4,060		143	10,000
Private drawings			9,600			9,000			9,600
Margin for reinvestments/tax			10,000			–4,940			400

on many farms where appropriate adjustments have not been made for the cut in milk quota.

RENT EQUIVALENT

The term 'profit before rent and finance charges' has purposely been used in the financial illustration given in Table 4.1 so as to allow the introduction of the term rent equivalent. One of the limitations of most comparative accounts data is the treatment of all farmers as 'tenants' even though 70% of land is now owner-occupied. Net farm income and management and investment income are useful, particularly from an academic point of view, in the assessment of management efficiency but on the farm the main criterion is the profit remaining after meeting the 'actual rent equivalent'.

The 'actual rent equivalent' (ie rent actually paid plus finance charges), does not often equate to the 'true rental value' of the holding. It is more likely to reflect the time the farmer has been on the holding and how recently he has purchased additional land or carried out improvements. Very often one finds that farms with poor buildings and equipment have high actual rent and finance charges whereas farms with completely modern set-ups have little or no rent or finance charges to pay.

The current (1994) value of the buildings and fixed equipment on a well-equipped dairy farm is likely to be in the region of £1,600 per cow but may only have cost £400 per cow. If a 'rental value' is charged based on the value of these buildings as well as the land then the resulting figure is likely to be in excess of £160 per cow or more than £400 per hectare compared to actual rents in the region of £120 to £200 per hectare.

It is usually assumed that the 'rent not paid' for owner-occupied land can be used to service finance charges on borrowed capital. This is often an erroneous assumption because the 'rent' is frequently required to service property repairs and ownership expenses. In particular it is required to service improvements. This point needs to be carefully borne in mind by anyone contemplating starting farming or acquiring additional land.

LABOUR COSTS

The Agricultural Wages Board determines annually the wages paid to agricultural workers. Statutory rates are determined for ordinary workers, craftsmen, Grade II workers and Grade I workers. The rates paid to craftsmen, Grade II workers and Grade I workers respectively are 115%, 125% and 135% of that paid to the ordinary farmworker.

Mention was made in Chapter 1 of the tendency for farm wage rates to rise much more quickly than the price of milk, using data going back to the 1950s. Table 4.2 shows that the recent improvement in the milk price has stopped this trend for the time being. This may explain in part why the independent members of the Agricultural Wages Board have given a rise of 4.9% to agricultural workers in 1994, compared to an inflation rate of only 2.5%. It also

Table 4.2 Labour costs

Per week Year ended	*Craftsman's basic wage 1983–84 per week*	*Milk price per litre*	*No. litres per wage*	*Index = 100*
March 1979	47.30	10.43	453	73
March 1984	91.08	14.67	620	100
March 1989	119.83	17.66	678	109
March 1990	128.82	18.93	680	110
March 1991	140.41	18.94	741	119
March 1992	148.84	18.94	785	126
March 1993	154.80	20.20*	766	123
March 1994	159.06	21.70*	732	118

* Estimates

helps to explain why dairy farming is much more profitable at the present time than it was a few years ago.

POWER AND MACHINERY COSTS

A reminder is needed about machinery and equipment depreciation and the points discussed in Chapter 3, in relation to inflation. In most farm accounts depreciation is determined on the historic costs basis, and consequently the depreciation charges shown need to be at least doubled to arrive at a realistic figure.

Power and machinery costs, like labour costs, have risen more rapidly than milk prices and control of these costs is another major challenge to dairy farmers. A feature of the development of dairy farming during the past 20 years has been the increasing size and sophistication of tractors and forage harvesting equipment. Consequently, dairy farms often have very high power and machinery costs as well as high labour costs relative to other farming systems.

One would expect profitable dairy farms to produce much higher gross margins per £100 labour, and per £100 labour and machinery compared to the average farm, but this is not always the case. This is illustrated by Table 4.3, taken from the report produced by Manchester University in March 1993 on the results for 1991/92. The more profitable farms (selected in relation to their profitability per hectare) have a management and investment income of £486 per hectare, 60% more than the average of £303. The gross margin per £100 labour, and per £100 labour and machinery however are not much more than average.

VARIATIONS IN PROFITABILITY BETWEEN FARMS

Table 4.3* illustrates the enormous variation found in the profitability between farms when selected according to the profit per hectare. The high profit farms are only 2.4 hectares larger than the average farm but keep 24.3 more dairy cows than average. The stocking rate on the high profit farm is much better than average (see Table 4.4).

The high stocking rate in this instance is reflected in higher forage costs both per hectare and per livestock units (see Table 4.5) but often this is not the case. The high profit farms often produce more from the same forage inputs and this ability is a key to management success.

The high profit farms produce much more milk than the average farm due to a combination of more cows and a higher yield per cow (see Table 4.6), and this is the key to their high profit per hectare, ie they produce more milk and probably have more milk quota than average. The high profit farms produced 24% more milk than the average farm and this has been reflected in an increase in the management and investment income of 65%.

MILK PRICE RELATIVE TO FEED COST

Reference has already been made to the fact that the milk price has improved very significantly in recent years, relative to feed costs. How much this has changed can be assessed by looking at the data shown in Table 4.7 and Figure 4.2. In the 1970s, the average price of milk per litre was only 80% of the price of concentrates per kilogram, but by 1993/94 the price of milk had risen to nearly 150% of the price of a tonne of concentrates.

Contrast this with the information shown in Table 4.8, taken from various editions of the management handbook produced by John Nix. This shows tractor costs per hour rising from £3.33 in 1981, to £7.63 in 1993, and during the same period the cost of buying a tractor has increased from £9,500 to £17,000. During the same period the cost of forage harvesting grass has increased from £77 to £105 per hectare. These figures underline the point already made that dairy farmers today need to look very carefully at their fixed costs, particularly those in relation to silage making.

* Table 4.3 and the remaining tables referred to in this chapter are shown on pages 51–54.

Figure 4.2 Milk price to feed cost ratio

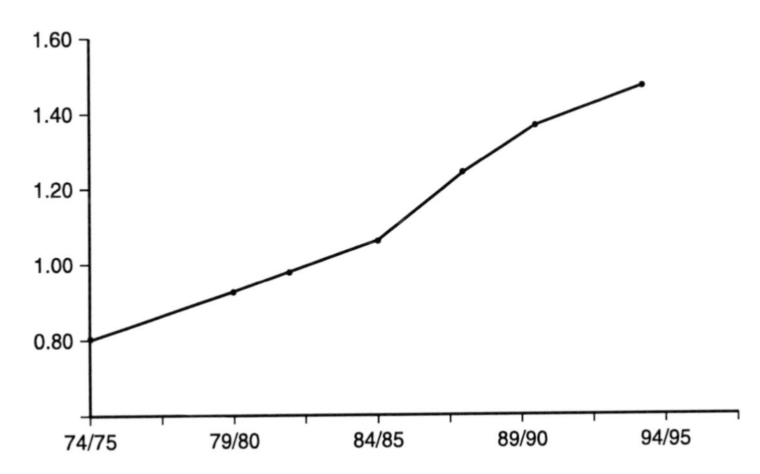

Spending large sums of money was easy to justify in the 1970s and early 1980s, but is much more difficult to justify today. It is salutary to note that concentrate prices today would be in the region of £240 per tonne if, in real terms, they were the same as they were approximately a decade ago.

Table 4.3 Financial performance—lowland dairy farms 1991/92

	Average farm	*High profit farm*
Farm size (hectares)	78.7	81.1
Dairy cow numbers	124.2	148.5
Other cattle numbers	66.5	61.2
	£ *per hectare*	£ *per hectare*
Gross margin excluding BLA*	1,347	1,633
Paid labour	173	230
Unpaid labour	191	190
Total labour	364	420
Power and machinery	367	426
Labour and machinery costs	731	846
Rent and rates	165	161
Repairs and maintenance	39	32
General overheads	109	108
	1,044	1,147
Management investment income	303	486
Gross margin per £100 labour	370	388
Gross margin per £100 labour and machinery	184	193

* Breeding and livestock appreciation

Table 4.4 Stocking rate

	Average farm No. livestock units	*High profit farm No. livestock units*
Dairy cows	124.2 × 1 = 124.2	148.5 × 1 = 148.5
Other cattle	66.5 × 0.5 = 33.2	61.2 × 0.5 = 30.6
	157.4	179.1
No. hectares	78.7	81.1
Livestock units per hectare	2.0	2.21

Table 4.5 Forage costs

	Average farm £ per hectare	*High profit farm £ per hectare*
Seeds	13	18
Fertilisers	110	126
Sprays	16	24
Other crop costs	23	38
	162	206
Livestock units per hectare	2.0	2.21
Forage costs per livestock unit	81	93

Table 4.6 Milk production profitability

	Average farm		*High profit farm*	
Size of farm (hectares)	78.7		81.1	
Number of dairy cows	124.2		148.5	
Yield per cow (litres)	5,627		5,849	
Total production (litres)	698,873		868,576	
Sale price per litre (p)	19.15		19.50	
	£	*Per litre*	£	*Per litre*
Total milk sales	133,834	19.15	169,372	19.50
Cattle output	19,474	2.79	18,617	2.14
	153,308	21.94	187,989	21.64
Other output	12,749	1.82	12,084	1.39
GROSS OUTPUT	166,057	23.76	200,073	23.03
Variable costs:				
Feed	37,540	5.37	40,631	4.67
Vet and med.	3,699	0.53	4,055	0.47
Sundry livestock costs	6,060	0.87	6,245	0.72
Forage costs	12,749	1.82	16,706	1.93
	60,048	8.59	67,637	7.79
GROSS MARGIN	106,009	15.17	132,436	15.24
Paid labour	13,615	1.95	18,653	2.15
Unpaid family/labour	15,032	2.15	15,408	1.77
Rent and rates	12,986	1.86	13,057	1.50
Machinery	28,883	4.13	34,549	3.98
Repairs and maintenance	3,069	0.44	2,595	0.30
General overheads	8,578	1.23	8,759	1.01
	82,163	11.76	93,021	10.71
MANAGEMENT & INVEST. INCOME	23,846	3.41	39,415	4.53
PLUS labour, farmer and spouse	12,434		12,408	
NET FARM INCOME	36,280		51,823	

Table 4.7 Milk price: feed cost

Year	*Milk price (pence per litre)*	*Concentrate cost (pence per kg)*	*Milk price to concentrate cut-rate*
1974–79	6.34	7.82	0.81
1979–80	11.56	12.91	0.90
1981–82	13.73	14.31	0.96
1984–85	14.57	14.00	1.04
1987–88	16.30	13.40	1.22
1990–91	18.94	14.10	1.34
1991–92	19.72	13.85	1.42
1993–94 (est.)	21.50	14.50	1.48

Table 4.8 Tractor costs and forage harvesting costs

Year	*1981*	*1985*	*1989*	*1993*
	£	£	£	£
Tractor costs per hour*	3.33	5.26	6.20	7.63
Tractor purchase price	9,500	13,500	13,500	17,000
Forage harvesting (carting and ensiling grass) per hectare	77	90	107	105

* 75–87 hp

John Nix, *Farm Management Pocket Handbook.*

CHAPTER 5

Enterprise Efficiency and Profitability Factors

INTRODUCTION

Chapter 4 has shown that the profitability of a dairy farm depends upon the relationship between fixed costs and the gross margin. It has also shown how the profit depends on the efficiency with which the farm is organised and managed and has demonstrated the ability of the better farmers to produce much higher gross margins than others from a given level of fixed costs.

This control of fixed costs in relation to the gross margin achieved by the business is, in the author's opinion, the most important factor determining the profitability of the farm, but unfortunately, most commentators on dairy farming profitability are only involved with and only look at the margin over concentrates (MoC) and similar margins.

Many advisers to farmers, for example, know what the feed and forage costs per litre should be on a farm, and can write a thesis on the factors that affect the MoC per litre without once mentioning the effect these might have on the fixed costs.

GROSS MARGIN PER COW AND PER HECTARE

The gross margin achieved per cow is the best way of measuring the efficiency with which a dairy cow is managed but grassland enthusiasts tend to see cows as converters of grass to money and prefer to see gross margin per hectare treated as the most important measure. Gross margin per cow, however, is a much better measure of return on capital.

The various factors that influence the gross margin per cow and per hectare are illustrated by budget data taken from the *Farm Management Pocket Book* by John Nix (see Table 5.1, for the year 1993/94).

The first thing to note is that the budget data is shown at four yield and four stocking rate levels. The data is shown in this way to demonstrate the enormous effect these two factors, that is the yield and stocking rate, have on the gross margin per cow and per hectare.

At average stocking rates, 1.9 cows per hectare, the budget gross margin per cow varies from £703 to £855 according to the level of milk yield. At average milk yields (5,350 litres per cow) the gross margin per hectare varies from £1,265 to £1,880 according to stocking rate. The full range in the gross margin per hectare is from £1,170 to £2,115. This tends to show that in effect there is no such thing as an average dairy farmer, as the average results achieved per cow and per hectare can vary so widely.

It is interesting to note that the budget data does not show any change in the forage costs according to the change in the milk yield per cow for a given stocking rate. Increased forage costs, however, are shown per cow as the stocking rate increases. Forage costs per hectare are assumed to be £200 at the very high stocking rate level but only £105.60 at the low stocking rate level.

One could perhaps question the assumption that forage costs do not change as the yield per cow increases. As the yield of the cow rises the amount of concentrates fed tends to rise and the quantity of forage such as home-grown silage tend to decrease. In other words, an increase in the yield and margin per cow is often associated with a lowering of the forage costs per cow.

This substitution of forage costs for concentrates is at the heart of the debate in dairy farming which has continued unabated for the past 40 years. The author remembers quite clearly a debate at Penistone, West Yorkshire, in the 1950s between the late Ken Russell from the Royal Agricultural College on the one hand, arguing for high yields, and ICI on the other arguing that cows should be fed more grass silage at the expense of concentrates with, if necessary, lower yields per cow.

This high yield, however, has to be achieved with proper control of the cost of concentrates. The main objective is to achieve a high margin, not a high yield regardless of costs. The budget data shows an increase in the feed costs as the yield rises, but it is assumed that this is controlled so that the MoC rises from £822 per cow to £989 as the yield increases from 4,850 litres to 6,350 litres. It is when one turns to concentrate costs that the problems involved in interpreting financial results begin, and this will be discussed in more detail later.

This control of concentrate costs in relation to milk yield is difficult to achieve and to assess. Basically, what a farmer has to resolve is how far he can go in the substitution of lower-cost home-grown food for more expensive concentrates without sacrificing yield. How far he can go and the economic consequences of these decisions are central to the controversy that centres round the feeding of dairy cows. The problems arise because one is trying to assess two factors at the same time, ie the efficiency of management of the dairy cow and the efficiency of management of grassland. Both have an enormous effect on the profitability of dairy farms. More effective grassland management explains in part why the better farms achieve higher yields per cow with no greater inputs of concentrate, forage or other costs per cow, all common features of surveys.

Improved grassland management, however, is not the full explanation. Some farmers achieve higher yields than others from a given input of concentrates because these are fed at the right time. They also achieve higher yields per cow because they have better cows and have a better standard of husbandry.

CHANGES IN COSTS AND PRICES

How rapidly costs and prices can change due to factors outside the dairy farmer's control is illustrated by comparing Table 5.2 to Table 5.1.

The data shown in 5.2 is taken from the 1994 handbook by John Nix, which has been published since the writing of this book commenced. The budget price for milk is 22.5p per litre for 1994, compared to only 20.25p in the previous year. The cost of concentrates is very much the same as in the previous year, ie £140 compared to £137.50. The so-called average farmer is assumed to have a yield of 5,500 litres compared to 5,350 a year ago.

The net effect of the above changes is that the budget gross margin before forage costs for the average farmer in 1994 is £961, compared to a budget for the very high yield farmer in the previous year, of only £926.

These budget figures demonstrate the significant improvement in profitability that is taking place in dairy farming at the present time, reflecting the freeing-up of the milk market and leaving the ERM in late 1992.

Table 5.1 Dairy cow gross margin budget data 1993–94

Performance level yield Litres	*Low 4,850*	*Average 5,350*	*High 5,850*	*Very high 6,350*
Milk price 20.25 p per litre	£	£	£	£
Concentrate cost £137.50 per tonne				
Milk sales per cow	982	1,083	1,185	1,286
Concentrate cost per cow	160	198	249	297
Margin over concentrates (A)	822	885	936	989
Calf output	100	100	100	100
Cull cows 22.5% £450.00	101	101	101	101
LESS cost of replacements 22.5% £750	–169	–169	–169	–169
Cattle output (B)	32	32	32	32
A+B	854	917	968	1,021
LESS				
Bedding		10		
Vet. and med.		28		
AI & recording fees		23		
Consumable dairy stores		24		
	80	85	90	95
GROSS MARGIN BEFORE FORAGE COSTS	774	832	878	926
Performance level—stocking rate:				
Low 1.65 cows per hectare				
Forage variable costs per cow	64	64	64	64
Gross margin per cow	710	768	814	862
Gross margin per hectare	1,170	1,265	1,345	1,420
Average 1.9 cows per hectare				
Forage variable costs per cow	71	71	71	71
Gross margin per cow	703	761	807	855
Gross margin per hectare	1,335	1,445	1,535	1,625
High 2.2 cows per hectare				
Forage variable costs per cow	75	75	75	75
Gross margin per cow	699	757	803	851
Gross margin per hectare	1,540	1,665	1,765	1,870
Very high 2.5 cows per hectare				
Forage variable costs per cow	80	80	80	80
Gross margin per cow	694	752	798	846
Gross margin per hectare	1,735	1,880	1,995	2,115

John Nix, *Farm Management Pocket Book*, 23rd edn.

Table 5.2 Dairy cow gross margin budget data 1994–95

Performance level yield *Litres*	*Low* *5,000*	*Average* *5,500*	*High* *6,000*	*Very high* *6,500*
Milk price 22.50p per litre	£	£	£	£
Concentrate cost £140 per tonne				
Milk sales per cow	1,125	1,237	1,350	1,462
Concentrate cost per cow	168	200	235	273
Margin over concentrates (A)	957	1,037	1,115	1,189
Calf output	110	110	110	110
Cull cows 22.5% £525.00	118	118	118	118
LESS cost of replacements 22.5% £950.00	–214	–214	–214	–214
Cattle output (B)	14	14	14	14
A+B	971	1,051	1,129	1,203
LESS				
Bedding		10		
Vet. and med.		31		
AI & recording fees		24		
Consumable dairy stores		25		
	85	90	95	100
GROSS MARGIN BEFORE FORAGE COSTS	886	961	1,034	1,103
Performance level—stocking rate:				
Low 1.65 cows per hectare				
Forage variable costs per cow	64	64	64	64
Gross margin per cow	822	897	970	1,039
Gross margin per hectare	1,355	1,480	1,600	1,715
Average 1.9 cows per hectare				
Forage variable costs per cow	71	71	71	71
Gross margin per cow	815	890	963	1,032
Gross margin per hectare	1,550	1,690	1,830	1,960
High 2.2 cows per hectare				
Forage variable costs per cow	75	75	75	75
Gross margin per cow	811	886	959	1,028
Gross margin per hectare	1,785	1,950	2,110	2,260
Very high 2.5 cows per hectare				
Forage variable costs per cow	80	80	80	80
Gross margin per cow	806	881	954	1,023
Gross margin per hectare	2,015	2,200	2,385	2,557

John Nix, *Farm Management Pocket Book*, 24th edn.

These figures also underline the tremendous improvement in the price of milk relative to the cost of concentrate. The budget price of milk for 1994 at 22.5p is 140% of the cost of concentrates (16p per kilogram). Prior to the introduction of quotas the price of milk per litre was less than the cost of concentrates per kilogram and these figures, once again, demonstrate a theme of this book, namely that high feed costs are no longer the dairy farmer's main problem.

The budget cost of a replacement dairy heifer is shown in 1994 at £950, compared to only £750 a year earlier and the budget income from the sale of a cull cow is shown at £525 compared to £450 in 1993.

These budget figures for 1994 were produced in the summer of 1993 when the average price of a cull cow was in the region of £525 and the average price of a dairy heifer was in the region of £950. Market trends since then have resulted in cull cow prices falling slightly compared to a year ago. The actual price this year (1994) is likely to be in the region of £450, ie the same as the budget for 1993, whereas at the present time (May 1994) the cost of purchasing a replacement heifer is well in excess of £1,000.

No criticism is implied in drawing attention to these discrepancies. They simply demonstrate how difficult it is to forecast future trends for factors that are as volatile as dairy replacement prices and cull cow prices. By the time this book is read, these prices will almost certainly have changed again. The price of a replacement dairy heifer, for example, in say one year's time, ie 1995, could again be back at a figure in the region of £950 per head, reflecting the effects of supply and demand and the probability that by then we shall tend to be producing well over quota on a national basis.

CATTLE OUTPUT

The data shown in Tables 5.1 and 5.2 has been set out in such a way as to allow the cattle output to be determined per cow. It has been set out in this way for two reasons:

1. It has become a much more important aspect of farm profitability since milk quotas were introduced, especially at times when the cost of leasing in quota is high.
2. It highlights the effect that the recent increase in the price of dairy replacement has on the cattle output per cow (see Tables 5.1 and 5.2).

The budget cattle output in 1994 is only £14 compared to £32 in 1993. In other words, a period of relative prosperity in dairy farming due to an improvement in the milk price and good calf prices is being offset in part by the high cost of purchased dairy replacements. However, farmers who rear and sell replacements surplus to requirements will find this a benefit, not a problem.

The cattle output is the difference between, on the one hand, the value of the calves produced by the dairy cow, and on the other the cost of maintaining the dairy herd, often referred to as herd depreciation. We now need to look at these two items in more detail.

CALF OUTPUT

The calf output for a given herd is determined by the number of calves born multiplied by the price received per head after allowing for mortality.

The significance of the number of calves reared per annum from the dairy herd should not be underestimated. On many dairy farms, particularly those where replacements are purchased, the number of calves reared tends to be significantly less than the number of cows in the herd. As mentioned elsewhere in this book, the number of calves reared depends on the number of dairy heifers that calve down as well as the number of cows in the herd. Such factors as the calving index are also important. On a well-managed dairy farm a detailed record is kept of the number of calves reared so that a rolling 12 month calf output can be monitored, as well as a rolling 12 months' total MoC.

It is the keeping of these records that has highlighted to the author the fact that the number of calves reared on farms where replacements are purchased is roughly equal to the number of cows in the herd *less* the number of replacements. For example, in a 100 cow dairy herd with a 20% replacement rate the number reared is in the region of 75 to 80, not 95 to 100. This means that the figures shown in Tables 5.1 and 5.2 for calf output are probably overestimated as they are based on the assumption that just under one calf is born for every cow in the herd, based on a calving index of 385 days. This point is referred to when considering the place of dairy replacements on the farm and is mentioned again here to draw attention to an error that is often made in the keeping of records and accounts for dairy farmers.

The price received per calf is largely determined by factors outside the dairy farmer's control. It is influenced in particular by the profitability or otherwise of beef production at the time of calf disposal, and whether or not there is good export trade for calves to Europe.

Calf values are subject to wide fluctuation in price from year to year and within years and exert a considerable *external* influence on the profitability of milk production. There is a tendency for prices to be low in the autumn when supply tends to be at its height and beef prices low, and high in May, when supply tends to be low and beef prices are high, but this trend is not always consistent. It can be completely masked or exaggerated by a temporary closure of the European market or by the sudden upsurge in the demand for calves resulting from an improvement in beef cattle prices.

To have calves available for sale when the market is in short supply is not easy to arrange but for those who take the trouble the rewards can be significant. (General maxim – keep if prices are low in the autumn to sell in the spring but sell in the autumn if prices are good.)

Most differences in calf value not due to seasonal or supply and demand factors are due to differences between breeds. The low price received for Ayrshire steer calves has been a major reason for the gradual replacement of many Ayrshire cows by Friesians and explains the reluctance of many British Friesian breeders to move over to the now more popular Holstein Friesians or pure Holstein.

The Channel Island breeds also suffer from the severe disadvantage of producing calves with a low value. Criticism is now levelled at the poor calves produced by the Holstein cow. Unlike the Ayrshire, however, the Holstein produces a higher yield than the Friesian and this will probably more than compensate for the disadvantage of the relatively low price of its calf.

The significance of the role in British agriculture of beef from the dairy herd is not to be underestimated. The best way in which most farmers can take advantage of this is by crossing a proportion of their cows to a beef bull. Traditionally this would have been to a Hereford, or possibly Aberdeen Angus, but calves from Continental breeds now command higher premiums. These premiums have to be assessed in the light of the possible adverse effect of a difficult birth of a large calf on the subsequent lactation yield.

A further advantage gained by using a beef bull is the opportunity this provides to run a bull with the dairy herd towards the end of the calving season. This helps to improve the calving interval, increases the number of calves born per annum, and improves the yield per cow, as well as providing calves of a higher sale value.

A note of caution is required regarding the extensive use of beef bulls which produce short-term gains in the way of increased calf output but can reduce the potential long-term development of the herd especially from a genetic point of view. Tremendous advances are being made at the present time in the development of genetics and there are a substantial number of very well-bred young Holstein bulls available, most of which are being slaughtered as young calves. These young bulls are probably as well bred, if not better bred, than some of the bulls available from semen companies. It is generally accepted that the best genetics in the dairy herd are usually in the youngstock and it would seem that now is a time for a progressive dairy farmer to contemplate using a young Holstein Friesian bull on his dairy heifers, rather than the traditional beef cross. This should not be regarded as a substitute for the heifers being bred from the dairy cows in the herd, but as an additional source of dairy replacements.

HERD DEPRECIATION (OR HERD MAINTENANCE COSTS)

This is shown in John Nix's budget fixtures for 1993 at the same figure of £67.50 per cow irrespective of yield.

To arrive at this figure the following assumptions are made:

1. The purchase price of a down-calved heifer (or the market value of a home-reared heifer) is £750.
2. The sale value of cull cows including casualties is £450.
3. The replacement rate is 22.5%, ie for a herd of 80 cows the number culled and replaced each year is 80 × 22.5% = 18.
   Given the above assumption, the herd maintenance cost is:
   £750 – £450 = £300 × 22.5% = £67.50 per cow.

Clearly it can be seen that the herd maintenance cost depends on the proportion of the herd that needs to be culled, and the difference between the purchase price of replacements and the average price received for cull cows. Most farmers rear their own replacements so the estimated herd maintenance cost is very dependent on the transfer value of the down-calved heifer.

It is very difficult when calculating the herd depreciation to take into account the improvement in the quality of the herd. It is normal practice to value all the cows at the same value at the beginning and end of the year irrespective of the change in the quality of the herd, but from time to time the appreciation of the value of the herd due to market forces is taken into account. There are times when a decision can and should be made to take action that will artificially increase the herd depreciation cost to the long-term benefit of the dairy herd. For example, it could be prudent just before the end of a particularly good year to replace the potential culls in the herd to ensure a continued favourable trend in the coming year. One of the case history farms referred to in Chapter 16 did exactly this in February/March 1994, culling 15 cows and replacing these with 15 newly calved cows.

Most farmers in their costings put conservative values on the heifers they transfer into the herd. This leads to an underestimation of the herd maintenance cost and an underestimation of the gross margin produced by the dairy replacement enterprise. The effect of this conservative value on the average dairy cow gross margin tends to be modest in percentage terms but it often has a very significant effect on the dairy replacement gross margin.

Returning to cull cow prices, these, like calf prices, tend to be high in the spring and early summer but low in the autumn and early winter when most cows are culled. Cull cow prices, like calf prices, are very dependent on factors outside the dairy farmer's control. Again, like calf prices, they are very much influenced by the breed of cow, the large cow such as the Holstein Friesian making much more money than the smaller Ayrshire or Channel Island cow. Earlier, emphasis was placed on the value of the calf as a major reason for the present popularity of the Holstein Friesian breed. With this, one should couple the high price received for the cull cow.

What can an individual farmer do about his cull cow prices? The simple quick answer is, very little, but he can try to avoid being placed in a position of having to sell cows in poor condition. The aim should be to sell cows in good condition and at periods of high prices. One of the advantages of the self-feed silage system, for example, is that cull cows tend to be in good condition due to their ability to put on weight on an ad lib silage diet. If cull cows are in poor condition, particularly in mid-winter, it often pays to keep them separately on silage and/or to keep them to sell them off grass in May/June when prices are high. The margins per month to be gained from this exercise, ie increase in value of the cow per month, can sometimes be as high as the margin over concentrates. Whether this should be done, however, does depend on the condition of the cows and the availability of feed supplies. It should also be noted that this is also one of those cases where doing the right thing for the profit of the farm as a whole can lead to a reduction in the margin per cow. The MoC per cow will be lower than it otherwise would have been if the cull cows are retained in the herd for a little longer.

This can also be true when we come to consider the other factor determining herd maintenance cost, ie the proportion of the herd culled per year. It is generally accepted that a good calving index is necessary to achieve good margins but this can be taken to extremes, for example by making a decision to cull any cow that is not in calf within 120 days of calving, irrespective of its age or previous milk yield. The cost of replacing this cow (ie the difference between its value as a cull and the price of a replacement) is likely to be at least £400 and this represents 4 to 5 months' MoC. The national replacement rate is in the region of 20–25%. Infertility or difficulty in getting in calf is a major reason for culling and in many instances this figure is unjustifiably high.

One final point needs to be made about culling rates. This determines the number of heifer replacements that need to be reared, which in turn can influence the number of cows that can be kept and/or the area of another enterprise such as cereals that can be grown. This indirect effect on the profit is often more significant than the direct effect on the herd maintenance cost. This is particularly true at the present time. A farmer with 25 dairy heifers due to calve this year and a requirement for only 20 can sell five on the open market and can command a premium over what he would receive for the sale of five cull cows in the region of £2,000 to £3,000.

Finally, as a rule of thumb, it is useful to note that calf output should exceed the herd maintenance cost on a well-run dairy farm by almost as much as the sundry variable costs, which means that in effect the gross margin is simply the MoC less forage costs.

APPRECIATION IN BREEDING LIVESTOCK UNIT VALUES

This was referred to in Chapter 3 when considering standard data from costings by Manchester University.

Dairy cow unit value appreciation in this instance is the amount by which a dairy cow has increased in value between the beginning and end of the financial year. In the long term it does not really represent a profit because the true value of the cow has simply been adjusted in monetary terms to take inflation into account.

The value of dairy cows however, like houses, does not tend to go up in a straight line but is subject to short-term fluctuation. The overall trend tends to be upwards but there can be wide variations around the mean. There has been a substantial increase in the value of dairy cows between 1992 and the present time (1994), but prices can go down as well as up, as was mentioned earlier in this chapter. These changes in the value of cows need to be kept quite separate from the calculation of the herd maintenance cost and should be shown in the accounts as a separate item, as discussed in Chapter 3.

The forward-looking dairy farmer, however, tries to bear these trends in mind when preparing his development plan for the business. Now (1994) is a good time to be selling stock surplus to requirements, and is a bad time to be planning an expansion of the dairy herd. In periods of prosperity in dairy farming it can take 15 to 18 months to finance the purchase of a dairy cow from the MoC. In periods of relatively low profitability it can take no more than 6 to 8 months. The appreciation in the value of dairy cows in the long term is a factor that needs to be taken into account when comparing dairy farming profits with arable farming profits. It is also something that needs to be very much borne in mind from a tax planning point of view as discussed in Chapter 14.

SUNDRY VARIABLE COSTS

These include bedding, veterinary services and medicines, AI and recording fees, and consumable stores.

The significance of bedding costs depends on the location of the farm. In the wetter all-grassland areas it is a significant item but in the Midlands and Eastern Counties it is a cost that can be largely ignored because bedding supplies are plentiful and cheap.

John Nix assumes that very little change takes place in these sundry costs as yields rise and this tends to be borne out by survey data. Care has to be taken to control these costs, like any other costs on the farm, but their control is not a crucial factor determining the whole farm profit.

SEMEN AND AI COSTS

A noticeable feature in the accounts seen for dairy farmers in recent years is the substantial funds now being spent on semen and AI fees. The author's attention was drawn to this by working out what it cost to produce a calf for taxation purposes, based on the parameters outlined in Chapter 14, in other words, the cost of AI/semen fees and supplementary feed fed to the cow during late pregnancy. It was thought that the cost of AI/semen would be in the region of £20 to £30 per calf reared, but in fact it came to no less than £50 per calf reared, including all beef calves born. The cost per dairy heifer calf reared was approaching £200.

There is tremendous interest in genetics at the present time and high sales pressure from semen/AI companies to invest large sums of money in AI/semen, as well as embryos. The prudent dairy farmer should examine this aspect of his business in more detail and consider in particular the keeping of a well-bred stock bull on his farm. Pedigree farmers have traditionally kept one or more well-bred stock bulls and the commercial producer interested in genetics should think seriously about following their example.

FORAGE COSTS

These represent the variable costs incurred to provide grazing and winter feed for the dairy cow and include seeds, fertilisers, sprays and crop sundry items, such as silage additives.

Contract charges for silage-making and grass keep are sometimes included as part of forage costs. They are variable costs but, for comparative purposes, it is much better if these items are included as part of fixed costs. Purchased forage feeds, such as hay, are also included, sometimes as part of forage, but it is much better if they are included under a heading separate from both concentrates and forage as 'other purchased or bulky feed'.

Traditionally, forage costs have been treated as part of the variable costs involved in running the dairy farm. The author's experience, however, in both managing a dairy farming business and providing consultancy services, has resulted in the view that forage costs should be treated as part of the fixed costs, as in practice they cannot be readily allocated between the individual livestock enterprises, and do not vary as a result of modest changes in the numbers of livestock kept on the farm.

The interpretation of forage costs data is one of the most difficult exercises in farm business management, as they are influenced by so many factors. When looking at the forage costs one has to try to take into account so many elements at the same time, such as the inherent quality of the land, the stocking rate, system of grassland conservations, the level of concentrate feeding and milk yield. In addition, one has to make this interpretation knowing that the forage costs have probably been allocated to the individual enterprises on a livestock unit basis and this allocation may be inaccurate. The first step in any analysis of forage cost data should therefore be to examine these in relation to the grazing livestock enterprises as a whole, *and then, and only then,* in relation to the individual enterprises.

A final point to remember is that many farmers' knowledge of the precise number of hectares they farm is hazy. The area of cash crops grown is usually known fairly accurately but the area of forage is often arrived at by simply deducting the area of cash crops from the total farm area, including buildings and roads. Beware, therefore, of any figures pertaining to stocking rates; these may or may not be accurate! The same point applies to forage costs; accurate records are often kept of fertilisers used on cash crops and the 'difference', ie total less these, is then charged to grassland!

FORAGE UTILISATION EFFICIENCY

Attention has already been drawn to the fact that some farmers can produce more milk per cow than others at little or no more feed cost, and also achieve this at higher stocking rates. This is due, in no small

Table 5.3 Energy production from forage (GJ)

	*Top 25%**	*Bottom 25%**
	GJ per cow	*GJ per cow*
Energy requirements:		
Maintenance plus calf production	25.7	25.7
Milk	29.6	24.7
Total	55.3	50.4
LESS		
Energy provided by concentrates and purchased bulk feed	21.3	19.7
Energy (by difference) provided by forage	34.0	30.7
	Cows/ha	*Cows/ha*
Stocking rate	2.41	1.50
	GJ per ha	*GJ per ha*
Energy (by difference) provided by forage	81.9	46.1

* Selected according to gross margin produced per hectare.
MMB Report No. 33, *An Analysis of FMS Costed Farms 1981–2.*

measure, to their ability to utilise grassland and other forage crops more effectively than the average farmer.

This ability to produce more from forage is illustrated in Table 5.3. This compares the energy produced from forage crops by the top 25% of farms to that produced by the bottom 25%, selected according to gross margin per hectare and expressed as GJ of ME per hectare. The theoretical energy requirements in terms of ME have been calculated for maintenance, calf and milk production, and from this has been deducted the amount provided by concentrates and other purchased feed. The difference gives the amount produced by forage.

There are disadvantages with this calculation as it is assumed that the use of concentrates is 100% efficient, and it is also a somewhat difficult calculation to make. However, it is a good indicator of technical efficiency and is closely related to profitability.

More recently, the tendency has been to aim to measure forage utilisation efficiency by calculating the number of litres of milk produced by forage. This is also an excellent way of measuring the efficiency with which forage is used. There is, however, a tendency to place too much emphasis on the yield that is achieved from forage at the expense of the yield that is achieved per cow.

MARGIN OVER FEED AND FERTILISER AS A MEASURE OF EFFICIENCY

In dairy farming, the margin over feed and fertiliser is used to measure two things at the same time, ie the efficiency of dairy cow management and the efficiency of grassland management.

Let us just look for a moment at an arable farmer who keeps pigs and grows cereals to feed his pigs. He works out a gross margin from growing cereals and a separate gross margin for keeping pigs, but if he did not keep a record of barley yields he would have to be content with a pig gross margin per hectare. The latter would be of very little use to him, as a low figure could be due either to a poor performance by the crops or a poor performance by the pigs. He therefore keeps records of yields so that the two can be separated.

We have a similar problem on the dairy farm but because it has always been accepted that we do so, we try to assess the performance of both the cows and the grassland with one figure. At the end, we do not really know whether it is the cows that are good or the grassland management that is good, or poor, as the case may be.

If we take an extreme case, the place of beef cattle on the farm, it is generally accepted that these produce a low gross margin per hectare. If, however, we take calf-rearing and express this on a per hectare basis we find it is very high, but this does not mean that calf-rearing is a good way of utilising grassland. The gross margin is high simply due to the fact that calves use little land. The highest gross margin per hectare is produced by an enterprise that does not use any land. This explains in part why high stocking rates are nearly always associated with high gross margins per hectare, as illustrated at the start of this chapter, with information shown in Tables 5.1 and 5.2.

This high gross margin per hectare also tends to have a very high correlation with the profit made per hectare. The fixed costs, ie rent, labour and farm machinery, fall expressed as a cost per litre, as the stocking rate rises. This reduction in the fixed cost per litre in today's economic environment more than makes up for any increase in the feed costs per litre.

There is always the inherent danger that an increase in the stocking rate will lead to a shortage of home-grown feed and to an increase in the amount of purchased feed. This had serious implications when the prices of concentrates and silage substitutes were high in real terms. Today, however, they are not high in real terms so one of the ways of ensuring success in dairy farming at the present time is to have a high stocking rate and accept the risk of a shortfall in home-grown feed supplies, as these can be made up by purchased bulk foods or concentrates, as appropriate.

MILK FROM GRASS

Producing as much milk as possible from grass will continue to be a key economic factor as well as husbandry factor on most dairy farms in the West. On these farms, measures such as litres obtained from forage will continue to be of vital importance.

Farmers in the Midlands and South East, however, have access to other crops such as forage maize and whole-crop cereals, as discussed in Chapter 7. On these farms the forage maize and whole-crop cereals will in effect become their forage and in these instances it would be more appropriate to talk in terms of milk from forage, including home-grown cereals, instead of milk from grass.

Students, including the author, have been taught to cost cereals based on their opportunity cost and to charge them to the cows based on their potential sale price, but this is rather a pointless exercise if the wheat crop is being grown expressly as a means of feeding the dairy cows.

The author is now involved with several mainly-grass farmers who also grow a modest area of wheat. This wheat crop is now treated as part of the forage from a costings point of view, and the calculation of the margin over feed obtained per hectare. In effect, home-grown wheat is being substituted for silage. It is quite pointless to treat the wheat as a concentrate and then talk in terms of a reduction in the amount that is being produced from forage, when in fact the amount of milk being produced from home-grown foods is increasing.

MILK PRICE

Reference has already been made to the substantial increase in milk price between 1993 and 1994, based on prices that are likely to be paid/have been paid by the Milk Marketing Board.

As from 1 November 1994, the price will depend on to whom it is sold, as well as the factors that the Milk Marketing Board presently take into account in determining their milk price paid to farmers.

Most competitors to Milk Marque talk in terms of a price based on the expected market leader, that is Milk Marque, so the rest of this chapter is devoted to the factors that determine the milk price received from the Milk Marketing Board on the assumption that most of these will stay in place after vesting day.

The first point to make is that the milk price shown in Tables 5.1 and 5.2 is the price received by the producer after deducting transport costs, capital contribution and co-responsibility levy. In the future, transport costs will be shown as a separate item and producers will need to get used to looking at differing price structures. In practice, it is likely to be exceedingly difficult to compare the price paid by one purchaser to another.

The main factors determining the price paid by the MMB are:

1. Compositional quality
2. Channel Islands premiums
3. Antibiotics
4. Hygienic quality
5. Cell count
6. Seasonality

Compositional quality

An individual producer's price is determined by applying the constituent values, shown below, to the average of the compositional

Constituent	*1993* *Pence per 1%*	*1994* *Pence per 1%*
Butterfat	2.223	2.239
Protein	3.293	3.497

quality tests (fat and protein) taken on milk consigned by the individual producer.

Over recent years the Milk Marketing Board has put increasing emphasis on the price paid for protein relative to butterfat and this is now having implications in the breeding of dairy cows as well as in respect of the monthly milk cheque. This emphasis on protein reflects the increased sales of low butterfat, semi-skimmed milk and the importance of cheese and other high-protein products.

Premium for Channel Islands

Channel Island producers were paid 0.216 pence per litre in addition to the price based on constitutional values in 1993.

Antibiotics

It is most important to ensure that milk supplies do not fail the antibiotic test as the payment in case of a failure is only 1.00p per litre.

The MMB, however, does provide each producer with a free insurance scheme that automatically covers accidental contamination of his bulk tank, providing no claims have been made in the previous 6 months and providing contamination is reported before collection.

Hygienic quality

Payments in respect of the hygienic quality of milk were first introduced in October 1982. This payment is based on the total bacterial count (TBC) and the TBC level determines whether a supplier receives an addition or a deduction which is incorporated into the basic price as shown in Table 5.4 on page 74.

The penalty for Band C increases to 6.00p if there has been a similar problem in the previous 6 months and to 10.0p if it occurs again within a period of 6 months. In other words, the penalty for lack of hygiene is severe.

Cell count

The rates applicable for the year 1993/94 are as set out in Table 5.5.

Seasonality

Since 1 April 1991, a great deal of emphasis has been placed on changing the seasonality of milk production. The price received for milk produced in the months July, August, September and October has been increased significantly compared to that received in other months of the year, as can be seen from examination of Table 5.6. This change in the seasonality of payments has resulted in a substantial change in the calving pattern of the national dairy herd, with much more emphasis now being placed on calving cows in the summer months.

The Milk Marketing Board is aware that this trend, if taken too far, could lead to a shortage of milk at other times of the year. A variation to the pricing structure for 1994/95 has been produced and this is also shown in Table 5.6. This places less emphasis on the payment of premiums for the production of milk in the months July, August, September and October. The emphasis, however, is still very much on mid-summer milk production and in the author's opinion this price structure will lead to a shortfall in milk supplies during the winter months November, December, January and February. This is a problem that will need to be addressed by all purchasers of milk, including Milk Marque after vesting day.

At the present time there is no premium payment to producers for level delivery. A substantial proportion of the purchasers of milk require level delivery and it will be interesting to see how long it takes for this to be reflected in the price paid to producers.

Table 5.4 Total bacterial count

		Milk price deduction (pence per litre)
Band A	20,000	Nil
Band B	20,001–100,000	0.25
Band C	More than 100,000	2.00

Table 5.5 Cell count rates for 1993–94

3-month geometric mean cell count/ml	*Band*	*Milk price deduction (pence/litre)*
400,000 or less	1	Nil
400,001–500,000	2	0.5
500,001–1,000,000	3	1.0
Above 1,000,000	4	2.0

Table 5.6 Monthly average net prices paid to wholesale producers (England and Wales)

	Percentage of constitutional values	
	Three years 1992/93/94	*1994/95*
April	– 2.9	–10.0
May	–14.5	–12.5
June	–11.6	– 5.0
July	+21.6	+15.0
August	+30.3	+15.0
September	+25.7	+10.0
October	+21.3	+ 5.0
November	+ 2.6	*
December	+ 2.6	*
January	+ 2.6	*
February	+ 2.6	*
March	+ 2.6	*

* Data not published as this will be post vesting day.

CHAPTER 6

Dairy Replacements

ROLE IN THE WHOLE FARM ECONOMY

Attention was drawn in Chapter 2 to the three main relationships between enterprises: competitive, complementary and supplementary. Emphasis was placed on the need to select a combination of enterprises and a system of farming that is well balanced. Dairy replacements are now considered, with these principles in mind.

The profits made from rearing dairy replacements are nearly always considerably less than those made from dairy cows. This is true whether the return is measured per hectare, per £100 labour or per £100 capital, but the margin in favour of cows has been considerably reduced following the introduction of milk quotas. Prior to the introduction of quotas many dairy farmers were justifiably criticised by their advisers for keeping too many youngstock. There are instances now, however, where the opposite is the case, ie farmers are not keeping enough youngstock to make up for their shortfall in milk quota. As a general rule, the number of dairy replacements relative to the number of cows should be kept as low as possible. Too many dairy heifers relative to dairy cows is a common weakness on many dairy farms and is usually due to either a high rate of culling from the dairy herd and/or too long a period between birth and calving. On most well-managed dairy farms with adequate quota, heifer numbers should not exceed 70 to 80% of the dairy herd numbers. It is, however, important to distinguish between having a high *enforced* culling rate, due to poor infertility etc, with cows leaving the herd at low prices, and a *planned* high culling rate where cows leave the farm at relatively high prices, eg as newly-calved cows having completed four to five lactations, the aim being to make room for heifers of superior genetic merit.

Despite their low relative profitability there are very few dairy farms on which the rearing of dairy heifers is not justified and over two-thirds of dairy farms supply all their own replacements. Farms on which heifer rearing is not justified are likely to have most of the following characteristics:

- Land that is all suitable for either grazing by the dairy herd, conservations or for growing arable crops.
- A severe shortage of working capital.
- Adequate buildings and equipment to house enough cows to keep staff both fully employed and motivated without the interest of rearing replacements.
- A manager or farmer capable of buying heifers as good as those he could rear on the farm.
- Last, and most important, a substantial milk quota.

Dairy heifer rearing is justified on a farm where some or all of the above conditions are not met. Most farms have an area of permanent pasture that is too far away from the buildings to be grazed by dairy cows and/or cut for silage and this in itself often justifies heifer rearing. On many farms this area is much larger than that required by the 70 to 80% of dairy heifers relative to cows, referred to above. This leads to too many heifers being reared and to a lower level of profit per hectare. In this case thought needs to be given to having an autumn-calving herd so there are dry cows to utilise the grazing area during most of the summer.

The provision of winter feed for youngstock usually competes with the needs of the dairy herd but there is a place for youngstock in the sense that their requirements for winter food are not as critical. Alternative feeds such as straw can be used in times of shortage, leaving all the silage for the dairy herd. Youngstock can also act as scavengers utilising leftovers from the dairy herd.

There is also a place for dairy heifers in times of prosperity as a means of 'hiding profits' in closing valuations based on cost of production.

PROFITABILITY IN THE WHOLE FARM ECONOMY

As dairy heifer rearing is basically unprofitable, relative to milk production, the main objective should be to manage this enterprise in such a way as to make effective use of the resources not required by the dairy herd.

Two-year calving is necessary to get high gross margins per head from a youngstock enterprise but the achievement of this objective may not be feasible if building resources are limited and may not be desirable if outwintering is a feasible proposition. In this instance it may be more prudent to calve at two and a half years of age and adjust winter feed costs accordingly, treating the youngstock as 'scavengers'.

Although it is emphasised that the rearing of dairy heifers is relatively unprofitable, this degree of unprofitability is often overestimated. This is due to two basic errors in most costing systems. First, the calf born to the heifer is credited to the dairy herd, *not* to the heifer that produced it, and the newly-calved heifer is usually transferred to the dairy herd at too conservative a value relative to the cost of purchasing newly-calved replacements. Second, forage costs are usually charged to the youngstock on a livestock unit basis. This system results in the youngstock being overcharged as they graze the less productive pastures and normally require and obtain less conserved forage than their livestock unit equivalents would suggest. When assessing the contribution of your youngstock enterprise, therefore, please try to ensure that the allocation of forage acres and costs is done on the basis of grazing records and known levels of conserved feed utilisation, not on livestock units.

The effect of these two errors on the gross margin obtained from a youngstock enterprise is illustrated in Table 6.1.

The youngstock enterprise in Example A shows a gross margin of £11,588, or £193 per head. In example B this increases by £3,112 to £14,700, or £245 per head. This is achieved by adding £120 per head for the value of the calves born to the heifers and by assuming a more accurate allocation of the forage costs.

The contribution made by dairy youngstock to the overall profit is also very much influenced by the trend over the year in market values. Their contribution is not correctly assessed if this is not taken into account. For simplicity's sake it is assumed in Table 6.1 that the value of the heifers at the beginning and end of the year is the same. During the past two years (1992 to 1994), there has been an annual increase in the youngstock values of approximately £250 per head. This needs to be added to the output and gross margin figures shown in Table 6.1 to arrive at the true contribution made by the youngstock.

The next conventional step in the analysis of the dairy replacements is to calculate the area of land used by the replacement, again using the livestock basis, and then determine the gross margin per hectare and compare this to that achieved by the cows. On most farms this is rather a pointless exercise as there is usually no similarity between the areas of land occupied by the cows on the one hand and youngstock on the other. The cows occupy the best grazing acres and utilise the best conserved forage. The youngstock

Table 6.1 Two examples of dairy replacement gross margin

Average no. replacements			*Example A* 60					*Example B* 60
				£				£
Heifers on hand at end of year	60	@	£400	24,000	60	@	£400	24,000
Casualties	2				2			
Heifers transferred to dairy herd	22	@	£1,000	22,000	22	@	£1,000	22,000
Heifers culled	2			800	2			800
Calves born to heifers					20	@	£120	2,400
Sub-total	86			46,800	106			49,200
LESS calves transferred from dairy herd	26	@	£150	3,900	26	@	£150	3,900
PLUS heifers on hand at start of year	60	@	£400	24,000	60	@	£400	24,000
Sub-total	86			27,900	86			27,900
Gross output				18,900				21,300
Variable costs:	*£ per head*			£	*£ per head*			£
Feeding stuffs	70			4,200	70			4,200
Veterinary fees and medicines	5			300	5			300
Sundries	5			300	5			300
Forage costs	42			2,512	30			1,800
	122			7,312	110			6,600
GROSS MARGIN	193			11,588	245			14,700

Forage cost allocation			
(A) Based on livestock units:			
Livestock units:	*No.*	*Factor*	*No. units*
Heifers over 2 years	12	0.4	4.8
Heifers 1–2 years	24	0.6	14.4
Heifers under 1 year	24	0.8	19.2
Total heifers	60	1.8	38.4
Dairy cows	90	1.0	90.0
Total livestock units			128.4
Total forage costs			£8,400
Forage costs per livestock unit		8,400 ÷ 128.4 = £65.42	
Forage costs allocated to dairy heifers		38.4 × 65.42 = £2,512	
Forage costs allocated to dairy cows		90 × 65.42 = £5,888	
(B) Based on grassland records:			
Youngstock:	*Hectare*	*Acres*	£
Grazing	10	(25)	1,000
Silage	4	(10)	800
	14	(35)	1,800
Dairy cows:			
Grazing	20	(50)	3,000
Silage	18	(45)	3,600
	38	(95)	6,600
TOTAL FORAGE COSTS	52	(130)	8,400

occupy the worst grazing land and utilise whatever conserved forage is available after the needs of the dairy herd have been met.

A further arbitrary value used in Table 6.1 is the value of the dairy heifer calves transferred into the replacement unit. In academic terms, one could argue that the figure used should be the value of the beef heifer calf that would have been produced if the cow had been served to a beef bull plus any savings in inseminations fees.

Having stressed the perils of using conventional gross margin data to measure the contribution made by heifer rearing, we need to move on to discuss means of improving the efficiency with which they are reared and how the contribution to profits can be improved. Before doing so, we need to ask the question 'Is there any point in routine farm business analysis in continuing to separate the youngstock gross margin from that of the dairy cows?' So many arbitrary assumptions have to be met, it is debatable whether it is worthwhile.

AGE AT CALVING

Age at calving, as already mentioned, is a vital factor. A low age of calving needs to be achieved consistent with the dairy heifer enterprise fitting into the farm system after the needs of the dairy herd have been met.

We need to be aware, however, that lowering the age of calving does not lead to any significant direct benefit resulting from improvements in the profitability of the youngstock themselves. Most of the benefits tend to be indirect. Reducing the age of calving tends to reduce the direct profit contribution of a given number of youngstock as feed costs tend to rise by more than the interest and other costs saved in their rearing, in other words, the youngstock gross margin goes down.

Lowering the age of calving becomes much more significant when we look at the land required (see Table 6.2). Calving at 33 to 36 months increases the area of land required by approximately 50% compared to calving at 24 to 27 months

The most significant factor is the 50% saving that can be made in winter feed, or hay equivalent requirement. This land is released for another enterprise and the real or opportunity cost of calving at a later date is the profit that could be generated from this land by the alternative enterprise. The area of conservation land that can be saved by lowering the age of calving is in the region of 0.2 hectare per heifer reared. Devoted to cereals, this land would give a gross

Table 6.2 Age of calving and land requirement

Age at calving		*24–27 months*			*33–36 months*	
Forage requirements		*High nitrogen* (hectares)	*Low nitrogen*		*High nitrogen* (hectares)	*Low nitrogen*
Grazing:						
First summer		0.10	0.15		0.10	0.15
Second summer		0.20	0.30		0.15	0.22
Third summer					0.20	0.30
		0.30	0.45		0.45	0.67
Hay equivalent:	*(tonne)*			*(tonne)*		
First winter	0.6	0.13	0.20	0.50	0.11	0.17
Second winter	0.9	0.20	0.30	0.75	0.17	0.25
Third winter				1.00	0.22	0.33
	1.5	0.33	0.50	2.25	0.50	0.75
TOTAL		0.63	0.95		0.95	1.42

margin of £500 to £700 per hectare or £100 to £140 per heifer reared. The opportunity cost of the 50% saving in the area required for grazing is much less significant on farms where there is a substantial area of permanent grass not suitable for grazing by the dairy herd and where cereals cannot be grown. The saving may simply be a reduction in fertiliser costs resulting from the lower stocking density.

A second consequential benefit of earlier calving is the reduction in building requirements and the labour required to tend the stock in winter. It is difficult to quantify this saving because, once again, it will depend on the alternative use of the buildings and labour.

Finally, and most important, there is a substantial cash flow benefit from earlier calving. This is particularly important where capital is limited or expansion is planned or needed in dairy cow numbers.

The extra cash generated by bringing forward the date of calving by 12 months is in the region of £1,000, ie £900 MoC (5,000 litres at 18p) plus £100 for the calf. This more than compensates for the additional cash required to rear the heifer which is not likely to be more than £100.

In an emergency, cash can be generated by selling off heifers surplus to requirements. The oldest group of heifers can be sold.

The next group can then feed on a higher plane of nutrition to calve at 24 to 26 months of age instead of 36. This cash can then be reinvested in dairy cows to bring about a substantial improvement in profitability.

The sale of , say, 16 bulling heifers would fund the purchase of ten cows. In one year these should produce a gross margin of £900 per cow or an extra £9,000 compared to a gross margin from the 16 heifers, in the region of £4,000 (£250 per head), a net increase of £5,000.

The above calculation demonstrates the substantial increase in profitability that can be achieved by switching capital investment from dairy heifers to dairy cows, providing that milk quota is available. In this example it is assumed that it is available and that an emergency has occurred because production has fallen below quota.

If production is not below quota, the calculations are not quite so straightforward as additional milk quota would need to be leased in for the ten cows. Between 55,000 and 60,000 litres could be required, costing in the region of £4,000, reducing the advantage for the ten cows over the 16 cows to £1,000, not £5,000. This reinforces the point mentioned earlier about the need for more, not less, youngstock if the quota is limited.

ORGANISING MORE PROFITABLE REARING

Having looked at length at the profitability of replacements in terms of their place on the farm, we now turn to examine other ways in which their contribution can be increased:

1. By rearing heifers of intrinsically better value. The cost of rearing a heifer of low genetic potential is no less than one with a high potential. It is important, therefore, to take care over the breeding policy. This must be kept in perspective, however, as management and feeding have just as much effect if not more on yields and margins from dairy cows. The cost of nominated services per heifer eventually reared is higher than necessary on many farms.
2. By rearing heifers born at the time that favours early calving, ie heifers born in the autumn rather than in the spring. The late winter and spring-born calf is not large enough to utilise grazed grass effectively in its first year and is difficult to manage to calve at two years without incurring very substantial feed costs. The best way to achieve this objective in many herds is by using

a beef bull on the dairy herd from the end of February or March until the end of September. This may necessitate some sacrifices in genetic potential but in most instances this is outweighed by the next factor.

3. By rearing heifers born over as small a time-scale as possible so as to provide even bunches of cattle for rearing, preferably one bunch. This facilitates the adoption of a controlled rearing programme designed to bring the heifers to the correct service weight, which in the case of Friesians is 330 kg by the time they are 15 to 18 months of age (see Figure 6.1). One of the biggest problems in heifer rearing is the wide age range often found in groups of cattle running together. The objective is to try to adopt a similar strategy to that favoured by efficient '18 month beef' producers.
4. If possible, plan the main calving data so that it fits in with the needs of the rest of the farm as well as the dairy herd. The ideal time to start, from the dairy herdsperson's point of view, may be the beginning of September, but if cereals are important to the farm economy it may be better to delay the start until the end of September.
5. Organise a work routine and calf-rearing policy that fits in with the facilities available. Have a spare capacity to deal with surplus steer calves so these can be retained in times of surplus.

   Individual bucket-feeding of calves is still regarded by many as the best method for keeping both mortality and feed costs down, but labour availability may necessitate a more streamlined and mechanised feeding system, even if this does mean an increase in feed costs per head.
6. Aim at a liveweight of 180 kg before turn-out in April. Calves of this weight or more make effective use of grazed grass. Try to turn the calves out before there is abundant grass available to avoid digestion problems.
7. Plan the first summer grazing strategy for the young calves. Ideally this will include a move to clean aftermath in late July, possibly after second-cut silage, with the objective of minimising worm infestations and maintaining liveweight gain.
8. Introduce supplementary concentrate feeding in late summer, if necessary, to reach target weight at yarding. On light land some economy in winter feed and concentrate costs may be feasible by growing kale and/or stubble turnips.
9. Bring cattle indoors in good time and if necessary give supplementary feed to reach target service weight. If cattle have to be out-wintered do not adopt a 'cheap feed policy at all costs' attitude—feed to reach target weights.

Figure 6.1 Target growth curve

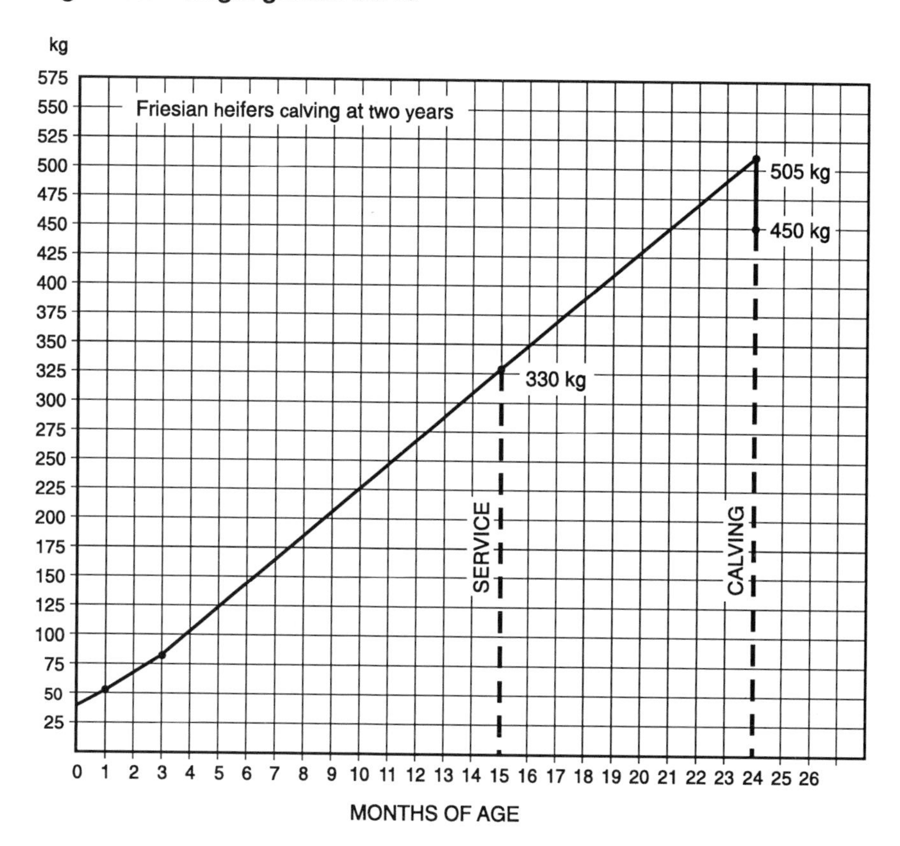

10. Plan the grazing policy for the following season with the same precision as you would that for the cows.
11. If there is a shortfall in dairy heifer numbers, rear beef heifers to make up this shortfall. These can be sold as and when necessary and will make good use of the resources that are available for youngstock rearing that would otherwise be wasted if these beef animals were not reared.
12. To sum up, the profitability of this enterprise is increased by giving as much thought as possible to its planning and day-to-day management. Dairy heifer rearing is a neglected enterprise and contributes less than it should to profits on nearly all farms.

REARING DAIRY HEIFERS FOR SALE

If we decide to rear heifers surplus to requirements for sale, we need to consider whether this is more profitable than an alternative enterprise. In considering this decision it is presumed that the needs of the dairy herd have already been met and that expansion of dairy cows is not justified.

The possible alternatives to dairy heifers will depend on the nature of the land available. If this is mainly permanent pasture then the comparison in profitability is most likely to be between dairy heifer rearing and beef cattle. In practice, there is relatively little difference in the profitability of these two alternatives and a decision can usually be made based on personal preference. If, however, the land can be used to grow cereals, and is eligible for Area Payments, it is most likely that a partial budget would show significantly in favour of growing cereals rather than keeping more youngstock. A decision to rear heifers in this case should be taken with the knowledge of the real cost to the farm economy of personal preference.

The case for rearing dairy heifers for sale rests largely on the price that can be obtained for the heifers. If a premium can be obtained over and above that achieved by the average farmer, it is more likely to be a profitable alternative to cereals and other livestock enterprises. This is most likely to be achieved by a pedigree herd and the place of heifer rearing in the pedigree herd is now considered in rather more detail.

PEDIGREE BREEDING AND REARING

At the outset, it is probably as well to define a pedigree herd. There are many commercial herds containing pedigree cattle but the main objective of the herds is to produce milk. A pedigree dairy herd is really a herd where the main purpose is to produce stock for sale and where milk, to some extent, is a by-product, although a very important one. The economy of these farms and the relationship between the profitability of milk production on the one hand and the production of heifers on the other is fundamentally different from that on the commercial farm.

Firstly, the cows being used to produce milk have a much higher opportunity cost, ie sale value, than that found on commercial farms. The average sale value will depend on the herd and its reputation. If it is a true pedigree herd as defined above, one would expect the sale value of heifers to be at least 150% of normal prices, ie £1,800 or more compared to £1,100 to £1,200. The sale value of elite cows in the herd is probably at least four times that of a typical commercial cow, ie in the region of £5,000 to £6,000, and could be even more.

It follows, therefore, that the profitability of such a farm is going to be very dependent on the price it can command for surplus stock, whether these be newly-calved heifers or cows having completed three or more lactations.

A good pedigree herd carries a large number of youngstock and sells very few cows as culls, probably less than 15% for two main reasons:

(a) Very few are culled because they are not in calf. The cows and their calves are too valuable to sell simply because they do not calve within a fixed date of say 420 days, and
(b) A large number of heifers enter the herd each year, probably in the region of 40%; consequently, most cows are 'culled' as down-calvers.

Pedigree dairy farms fared badly in the years immediately following the introduction of quotas, due to its effect on the value of surplus stock, but stock values recovered in 1993 owing to a more even balance between supply and demand.

Table 6.3 sets out the gross margin results that could be expected from a pedigree herd, alongside those that could be expected from a more typical commercial dairy herd. The gross margin of the pedigree herd is £19,700 more than that from the commercial herd. This is due to the substantial increase expected in surplus cow/heifer sales, made possible by the release of acres to feed the rearing of extra heifers, owing to the assumed higher yield of the pedigree herd, 6,666 litres per cow compared to 5,714.

REARING MORE DAIRY REPLACEMENTS

Many advisers stress the importance of rearing as few heifers as possible, placing particular emphasis on the opportunity-cost of the beef calves that could be produced if more cows are put to the beef bull. Serving an extra 20 cows to a beef bull, it is argued, will add an extra £100 per head or £2,000 to the calf income due to the higher value of the beef calves. It is forgotten, however, that only half of the beef calves will be bulls, problems could be experienced with calving, giving a total benefit of calf income of £1,000.

What matters is whether or not there are adequate resources already available on the farm to rear these extra 10 heifers, and

Table 6.3 Gross margin budget for a pedigree herd

	Pedigree herd			*Commercial herd*		
Average no. cows			120			140
Average no. youngstock			120			80
Farm size (ha)			80			80
Milk production (litres)			800,000			800,000
Yield per cow (litres)			6,666			5,714
Stocking rate (LUs per ha)			2.25			2.25
Calves born alive			140[a]			140[b]
Calves reared			60			40
			£			£
MoF[c]	800,000	@ 17.5p	140,000	800,000	@ 17.5p	140,000
Cull cows	15	@ £480	7,200	35	@ £500	17,500
Surplus cows/heifers	40	@ £1,000	40,000			Nil
Calf sales	60	@ £100	6,000	40	@ £100	4,000
	20	@ £150	3,000	60	@ £150	9,000
			196,200			170,500
LESS Livestock purchases			Nil?			Nil?
Youngstock feed	120	@ £75	9,000	80	@ £75	6,000
NET OUTPUT			187,200			164,500
LESS AI, semen and embryo	60	@ £80	4,800	40	@ £40	1,600
Vet. med. sunds. cows	120	@ £60	7,200	140	@ £60	8,400
Vet. med. sunds. heifers	120	@ £25	3,000	80	@ £25	2,000
Forage costs/hectare	80	@ £175	14,000	80	@ £175	14,000
			29,000			26,000
GROSS MARGIN			158,200			138,500

[a] 90 to cows, 50 to heifers. [b] 110 to cows, 30 to heifers. [c] Milk sales LESS feed cost.

what additional costs will be involved in their rearing. After two to two-and-a-half years the rearing of these additional 10 heifers will result in there being an extra 10 calves to sell, as well as an extra 10 cows to cull or sell surplus to requirements. In the vast majority of cases these will more than offset the opportunity-cost foregone of not having the beef calves income in the first year.

EFFECT OF REARING MORE HEIFERS ON OTHER PROFITABILITY FACTORS

It is often assumed that yield achieved by a particular herd of cows will be the same whether replacement rate is 20% or 25%. In practice, this is not the case as the fact that there are 25 heifers to enter the herd instead of 20 will probably result in the average number of cows being slightly higher than would otherwise have been the case, and this will give an increase in the total yield of the herd, probably in the region of that achieved by five heifers, ie 25,000 litres.

If this is not the case, it would probably result in cows being culled at a better time, and higher prices being received for culls. If there are adequate heifers coming forward, poor cows are not retained simply 'to make up numbers'; they are culled when they are worth, say, £500 rather than keeping them for an 'extra' lactation, which results in a casualty value at the end of their life of, say, £200. To conclude, it is stressed that farmers and their advisers should look carefully at the true place of dairy replacements on the farm, particularly at times when the cost of leasing in quota is high.

CHAPTER 7

Other Enterprises on the Dairy Farm

CAP REFORMS 1993

The role and profitability of 'other enterprises' on the dairy farm has been drastically changed by the introduction in 1993, as part of CAP Reform, of the Beef Special Premium Scheme for male beef cattle, the Suckler Cow Premium Scheme, the Sheep Annual Premium Scheme and the Arable Area Payments Scheme for combine crops and forage maize.

Details of these various schemes can be obtained from MAFF. The significance of the various schemes is commented on later when discussing each enterprise but the most important points are also summarised below.

The Beef Special Premium Scheme (BSPS) is restricted to 90 male cattle per farm and is subject to stocking-density limits. Animals qualify twice in their lifetime, once after reaching the age of 10 months and again after reaching the age of 23 months. Premium can be claimed on up to 90 animals in each category but, as already stated, claims are subject to stocking rate limits.

The stocking density limits, in livestock units per forage hectare, are:

1994	3.0
1995	2.5
1996 onwards	2.0

5,200 litres of milk quota is assumed to be equivalent to one livestock unit. Consequently, BSPS claims will not be accepted after 1996 on farms where the quota is in excess of 10,400 litres per forage hectare; in other words, a farm with 80 forage hectares will be able to claim BSPS payments only if its total milk quota is less than 832,000 litres.

The Suckler Cow Premium cannot be claimed by dairy farmers unless the milk quota is less than 116,450 litres (120,000 kg).

The Sheep Annual Premium Scheme is subject to a quota based on the number of breeding sheep on farms in 1991, and producers have been awarded individual quotas based on this year.

A dairy farmer who did not keep breeding sheep in 1991 can keep them on his farm without any penalties (unlike milk quotas where a levy would be payable), but cannot claim the annual premium. If he wishes he can either buy or lease quota to introduce or increase the number of breeding sheep on his farm. The purchase and sale of store lambs is not affected by this scheme as all the subsidy is now paid on the breeding ewes. The payment of the Sheep Annual Premium is *not* subject to stocking density limits but the presence of ewes on the farm is taken into account when calculating the entitlement to Beef Special Premiums.

A breeding ewe is assumed to be equivalent to 0.15 livestock units. If on the 80 hectare farm referred to earlier, claims were submitted for 100 ewes, this would be equivalent to 15 livestock units requiring 7.5 hectares. This reduces the eligible area for milk and beef production to 72.5 hectares and BSPS claims will only be accepted if the milk quota is not more than 754,000 litres ($72.5 \times 10{,}400$).

In practice, the above rules could result in dairy farms introducing breeding ewes, as they are not subject to stocking limits. No subsidies are paid in respect of beef heifers so these can be kept on intensive dairy farms and sold to beef producers as either finished cattle or as in-calf/down-calved suckler cows. Beef calves reared on the farm can be sold without claiming subsidy to beef producers who are able to submit claims. In practice, most of the potential subsidy is likely to be reflected in the price paid for the beef store.

Dairy farmers are also able to retain ewe-lambs for sale to qualifying sheep producers as shearlings so the effect on dairy farmers' profits of the introduction of the above scheme is not as serious as was first thought. Its impact on intensive bull beef systems is however significant and it is now difficult to justify on most intensive dairy farms.

The Arable Area Payments Scheme is restricted to land that grew eligible crops in the period 1986 to 1991. Land that was in grassland throughout this period does not qualify.* Cereals can, however, still be grown on non-eligible land and the area devoted to cereals can

* This is the case at the time of going to press but proposals have been made to change the rules and this could lead to some land that was in grass during this period being eligible.

be included in the forage area when calculating the entitlement to Beef Special Premiums.

Fifteen hectares of land has to be set aside for every 85 hectares on which arable payments are claimed, if the area claimed exceeds 15.51 hectares. No land, however, has to be set aside if the area claimed is less than 15.51 hectares, even if more than 15.51 are grown.

Forage maize qualifies for Arable Area Payments. There is now a separate base area for forage maize and this could lead to the payment being less than for cereals. This will occur if the actual area of cereals and maize grown in total is more than in the base reference year.

ECUs AND SUBSIDY PAYMENTS

The Beef Special Premium and Arable Area Payments are paid based on ECUs (European currency units):

Area Payment—cereals and maize	207.55 ECUs per hectare
Beef Special Premium	75 ECUs per head (1994)
Beef Special Premium	90 ECUs per head (1995)
Suckler Cow Premium	90 ECUs per head (1994)
Suckler Cow Premium	120 ECUs per head (1995)

The rates of premium in pounds sterling for beef are determined by the rate of exchange in January of each year; for Area Payments the date is July. In January 1993 the rate of exchange was 1 ECU = £0.93052; in July 1993 it was £0.948645.

These payments in pounds sterling are much higher than was expected when the schemes were devised, due to the decision taken in October 1992 by the government to leave the ERM (Exchange Rate Mechanism). The so-called 'Black Wednesday' was a 'golden day' for British farmers as the devaluation of the pound sterling led to a 19–20% increase in the value of the ECU, ie from approximately 78p to 93p.

In *certain cases* it will pay a dairy farmer to claim the Beef Special Premium and forego the Area Payment for cereals.

Returning to our 80 hectare dairy farm, let us suppose there is a milk quota of 676,000 litres and this requires 65 forage hectares, leaving 15 hectares devoted to cereals. The Area Payment claim for cereals would be 15 × 207.55 = 3113.25 ECU.

The number of ECUs that can be claimed in respect of beef cattle is shown in Table 7.1, based on calculations that take into account the stocking rate and the differing payments from year to year. The number of ECUs that can be claimed in respect of the 15 hectares in

Table 7.1 Beef Special Premium claim for 15 forage hectares

	Stocking rate allowed		*Forage hectares available*		*Livestock units*		*Max. no. of animals*	*Payment* per head (ECUs)*	*Total payment (ECUs)*
1994	3.0	×	15	=	45 ÷ 0.6	=	75	75	5,625
1995	2.5	×	15	=	37.5 ÷ 0.6	=	62.5	90	5,625
1996	2.0	×	15	=	30 ÷ 0.6	=	50	90	4,500

* If under 2 years of age at date of claim.

1994 and 1995 is 5,625 and in 1996 is 4,500, well above the number that can be claimed based on Area Payments for cereals.

Even in 1996 there is an advantage in favour of the beef claim of 1,386 ECUs and at an exchange rate of 94p, this is equivalent to approximately £1,300. Dairy farmers, therefore, need to get out their calculators and study the small print when submitting their IACS (Integrated Administration and Control System) and other claims. Note: Just to make it even more complicated, the above calculations have been made on the assumption there is no reduction in the beef premium payable due to the regional ceiling or the national quota for beef premium payments being exceeded. The regional ceiling in 1993 was exceeded by 24.7% and this means that a farmer who submitted a claim for 40 animals will receive payment for only 30 and the payment per ECU is in effect 70 pence, not 94 pence.

This scaling down reduces the advantage of beef over cereals but this is not likely to be so severe in future years.

Beef cattle 4,500 ECUs @ 70p = £3,150
Cereals 3,113 ECUs @ 94p = £2,926

There is a significant margin in favour of claiming the beef premium in 1994 and 1995 even at 70 pence, as the claim can be submitted in respect of 5,625 ECUs, and on balance it is considered that the advantage will continue to be in favour of beef cattle, not cereals.

A further point to consider before going on to discuss each individual enterprise in more detail is the quantity of milk quota deemed to be attached to a dairy farm for subsidy payment purposes. It is the quantity owned on 1 April. Quota purchased or leased after 1 April is not taken into account and this needs to be borne in mind when making management decisions over the timing of milk quota purchases and sales.

104,000 litres purchased on 1 May instead of 1 April increases the forage area available for Beef Special Premium claim by 10 hectares, and increases the potential subsidy payment by 3,750 ECUs in 1994 and 1995, equivalent at current rates of exchange to approximately £3,500.

The introduction of the stocking rate regulations is having a 'knock-on' effect on the amount farmers are prepared to pay to summer graze youngstock and beef cattle. Grazing prices are now being inflated by the additional payments that can be received if taking the additional land results in the appropriate stocking rate.

The discussion earlier about choosing between beef payments and cereal payments was made on the assumption that the area of the farm was limited.

If an additional 15 hectares were rented on a short-term basis, the area subsidy could be claimed for the cereals as well as the payments in respect of the beef cattle.

The area payment for cereals is £194 per hectare (£77 per acre) and this is in line with the amounts that have been paid in recent years for grass keep.

CEREALS ON THE DAIRY FARM

Cereals are found, and can be justified, on the majority of farms for the following reasons:

1. Properly managed, they produce a gross margin per hectare that is higher than that obtained from grazing livestock other than dairy cows unless the grazing livestock is very well managed, or particularly profitable due to an increase in youngstock prices, as for example between 1992 and 1994.
2. The working capital required to grow cereals is much less than that required for grazing livestock. Two to three hectares of cereals can be grown with no more working capital than that required for one hectare devoted to livestock.
3. They reduce the costs of bedding. (Note: When you assess the gross margin of a cereal enterprise on a dairy farm you should include the value of the straw as well as the value of the grain in the gross output.)
4. They make it easier to re-seed leys and facilitate the disposal of slurry and farmyard manure. Worn-out leys, for example, can be used to dispose of farmyard manure during the autumn before they are ploughed for wheat. Alternatively, slurry can be applied to the stubble during the winter before re-seeding in the spring.
5. A higher stocking rate can be adopted with less fear of being short of fodder in a drought year due to the availability of feeding straw.

6. Savings can be made in purchased concentrate costs by making use of home-grown grain. It is normal practice to charge grain fed to livestock at the price it would have realised if sold, but the real saving to the dairy farmer is what it would have cost to purchase cereals. This is some £4 to £6 per tonne more than the sale price due to transport costs, etc.
7. Cereals can be grown on the dairy farm with relatively low fertiliser costs because advantage can be taken of the build-up in fertility via leys and manure.
8. Area Payments can be obtained for growing cereals but they cannot be claimed for grass silage.

An indication of the gross margins produced by a wheat enterprise at average and above average levels of efficiency is shown in Table 7.2. These figures demonstrate the importance of high yields in obtaining good gross margins from cereals. To achieve these high gross margins it is necessary to drill the wheat crop early, that is, by the middle of October.

If winter wheat follows a ley this will need to be ploughed not later than the end of September and the utilisation strategy for the ley should be planned accordingly. This may conflict with the needs of the livestock enterprise such as late autumn grazing. The dairy enterprise manager may instinctively make the wrong decision in

Table 7.2 Wheat enterprise gross margins at 1994 price/cost levels

		Average per hectare (acre) £			*Above average per hectare (acre)* £	
Grain output	7.0 tonne @ £95	665	(266)	8.0 tonne @ £95	760	(307)
Straw	2.5 tonne @ £10	25	(10)	2.5 tonne @ £10	25	(10)
Area payment		185	(78)		185	(75)
		875	(354)		970	(392)
Variable costs:						
Seed		45	(18)		45	(18)
Fertiliser		65	(26)		65	(26)
Sprays		100	(41)		100	(41)
		210	(85)		210	(85)
GROSS MARGIN		665	(269)		760	(307)

this instance by automatically giving preference to the needs of the livestock, even to the extent of sacrificing not only yield but also the winter wheat crop in favour of spring barley. The opportunity cost of this decision to grow a spring barley crop instead of winter wheat is likely to be in the region of £165 per hectare (£665 per hectare wheat gross margin less £500 per hectare barley gross margin).

Question: Has he made the wrong decision to grow barley instead of wheat? From a gross margin point of view, he has. Ten acres of spring barley instead of winter wheat reduces the gross margin by £1,650 (10 × £165). This, however, could be less than the hidden benefits (from a costings point of view) of having land on which slurry could be spread during the winter months, and a grass ley that could be grazed by youngstock/dry cows until, say, the end of December, thereby making a saving in silage requirements. The answer will depend on the individual farm circumstance.

The discussion to date has centred round the gross margin contribution a cereal enterprise can make when introduced onto a dairy farm as an enterprise in its own right. A feed substitution case for cereals can also be made: in other words, it is more economical to devote some land to cereals to feed to cows than to grow grass to conserve as silage. This may not often be valid in the wetter far-western half of the country but it is often true in the midland and eastern counties. The case for grass in favour of cereals rests largely on its ability to produce high yields per hectare. Some idea of how high this yield has to be can be gauged by calculating the silage yield that has to be produced to give the same gross output as that attainable from wheat (see Table 7.3). Yields in excess of 35 tonne per hectare (14 tonne per acre) are required to equate with growing a crop of wheat if the silage is valued at £25 per tonne. At £20 per tonne yields are required in the region of 45 tonne per hectare (18 tonne per acre).

One must then turn to the question, how much does it cost to produce these forage yields compared to growing wheat? Variable costs are likely to be very similar, savings in seed and spray costs required to grow grass being offset by additional fertiliser requirements. Fixed costs are difficult to assess and will depend on the individual farm circumstances but it is fair to say that the harvesting costs of cereals are likely to be much less than those for silage.

Yields of silage in the region of 45 tonne per hectare (18 tonne per acre) are feasible but increasingly questions are now being asked as to the advisability of continuing to base dairy farming systems on the utilisation of large quantities of grass silage, particularly when a subsidy (Area Payment) can be claimed for cereals but not for grass silage.

Table 7.3 Output per hectare (acre) of wheat and silage

	Silage value *per tonne*	*Average wheat crop* *Per hectare (acre)*	*Above average wheat crop* *Per hectare (acre)*
Wheat output		£875 (354)	£970 (392)
Silage output	£25	35 (14.1) tonne	38.8 (15.7) tonne
	£20	43.7 (17.7) tonne	48.5 (24.2) tonne

FORAGE MAIZE AND WHOLE-CROP CEREALS

This questioning approach to feed substitutes and alternatives to grass silage has recently been adopted by farmers who have changed to complete diet feeding. The last ten years have seen a resurgence of interest in the growing of forage maize and this has been stimulated even further by the fact that forage maize is eligible for Area Payments, in effect, a subsidy to grow this crop instead of grass silage. Increasingly, dairy farmers and their advisers and research workers are becoming aware that grass silage is an expensive crop to grow and when made is often of doubtful/variable quality.

When milk quotas were first introduced the perceived wisdom was to aim to make more silage per cow, say 12 to 14 tonnes per cow, instead of 8 to 10, and to maximise the margin per litre. This perceived wisdom tended to emanate from farmers and advisers not fully conversant with the basic principles of dairy farm business management, ie people more concerned with technical as opposed to economic excellence.

The extent to which these attitudes have changed can be gauged by the title of the 1993 winter meeting of the British Grassland Society, 'The place for grass in land use systems', and some of the papers presented at this meeting. Papers presented included 'The impact of economic factors on crop and system options for the UK farmer', 'The experience of a farmer with a high reliance on maize silage', and 'The integration of whole crop wheat into grass silage diets'. There was also an excellent paper entitled 'The experience of a farmer relying on grass silage', as one would expect at a BGS meeting but the important point was that papers dealing with crops other than grass were discussed.

The area of forage maize has expanded for five main reasons:

1. When included in the diet it increases the protein percentage as well as the yield per cow.
2. It costs less to gr*ow than grass silage.
3. It makes good use of slurry and farmyard manure.
4. It is more environmentally friendly, as it produces less silage effluent than grass.
5. The eligibility for receipt of Area Payments.

The main disadvantages of forage maize are:

1. The geographical area over which it can be grown is restricted. Newer, earlier maturing varieties are needed to allow it to be grown farther north.
2. A separate silage pit is required and it is difficult to incorporate into the feeding system if this is based on self-feed silage.

It is considered that the advantages outweigh the disadvantages and the area devoted to this crop on dairy farms will continue to increase.

The droughts experienced in 1990 and 1991 led farmers and researchers to look at whole-crop cereals as an alternative to silage. The problem with whole-crop cereals is that they have a relatively low nutritional value due to the substantial proportion of 'straw' in the whole crop, relative to grain. It is an excellent feed for young-stock but has a limited role in the feeding of dairy cows.

Whole-crop cereals should be looked upon as an insurance crop in years when silage is in particularly short supply. In normal years it is almost always going to be more profitable to combine the crop, sell the grain and, if it is more profitable to do so, buy in an alternative forage such as brewers' grains.

This is particularly true if the opportunity is taken to increase the forage stocking rate by the growing of more cereals. In a dry year the grain will be sold to fund any shortfall in winter forage supplies. In a normal or wet year the grain can be sold and virtually all of the income realised will be extra profit. Growing cereals on a dairy farm removes virtually all the risks involved in having a high stocking rate. The important point to remember is that the cash surplus generated by the cereals will not be available every year, but it should be there in at least seven years out of ten.

These views on the place of whole-crop cereals on the dairy farm are given in the light of today's technology. It has to be said that to the dairy farmer it is a very attractive proposition that he should be able to conserve and harvest his wheat crop with similar machinery to that used for his grassland. 'Necessity is the mother of invention' and it is quite likely that during the next few years technology will have developed to such an extent that whole-crop cereals with a high nutritional value will become part of the ration on many dairy farms.

The whole-crop cereal, however, will tend to be seen as a substitute for concentrates, not for silage, and will be fed with the main objective of increasing yield, not simply reducing the cost of concentrates in a low yield system.

POTATOES

Potatoes produce a high gross margin per hectare and are an effective way of improving dairy farm profits, given the right climate and soil conditions. This is particularly true, for example, in certain parts of Devon and Cornwall, Pembrokeshire and Cheshire where the climate is ideal for early potatoes. Dairy cows and potatoes fit together very well on these farms because double cropping can be practised, for instance, early potatoes followed by kale or stubble turnips. Advantage can also be taken of the benefits farmyard manure has on potato yields, although these benefits are questioned by some experts.

Potatoes are a better choice than cereals—given that both crops can be grown well in the area—on dairy farms that are relatively over-staffed, for instance, the family farm needing to increase its output when the sons return home from college. Potatoes, however, are a specialist crop which require substantial amounts of capital per hectare. Whether they should be introduced will depend on the availability of capital and whether there are buildings on the farm than can be readily adapted for potatoes.

This availability, or otherwise, of buildings applies equally to cereals. There is usually a substantial amount of buildings on a dairy farm but a large proportion tend to be specialised. General-purpose buildings, however, can be used for both cattle and cash crops. Cereals, for example, can be stored until November and then sold to make room to house youngstock (although in due course, it should be noted, EU rules could lead to a ban on this common practice).

Note: If buildings and potato quota are not available it may be worthwhile to allow a potato grower to grow potatoes on a contract basis. Leys needing ploughing out can be chosen and the potatoes will provide an excellent means of disposing of manure and re-establishing a new ley. This practice is likely to become more attractive to a dairy farmer who has land *not* eligible for cereal Area Payments. The potato grower can grow potatoes on the dairy farm and more cereals on his own eligible land. There is a profit there to be made; it is up to the entrepreneur to exploit it!

SUGAR BEET

Sugar beet is found on farms where milk is produced but on such farms it is probably more a case of dairy cows being kept on an arable farm, rather than beet on the dairy farm. To produce a good profit, sugar beet needs to yield in excess of 35 tonne per hectare and these yields can only be achieved year-in-year-out on grade I and grade II land. Sugar beet, however, is grown on grade III land and it is on such farms that it is more likely to be associated with milk production.

Yields on such farms will tend to be below the national average and effective use will need to be made of sugar beet by-products to justify the crop continuing to be grown on the farm. The by-products take two forms: beet tops and beet pulp. The latter can be purchased by any dairy farmer through a merchant but the beet grower is able to purchase at a lower price, usually in the region of £4 to £6 per tonne. Although sugar beet pulp is very good feed for dairy cows, the advantage gained by being able to purchase at a lower price is not in itself very significant. Beet tops have considerably more potential value but simple means of using these effectively in the rations of dairy cows are difficult to devise. Some arable farms have made reasonable quality silage and dry cows can make good use of beet tops in late October to early December. They can also be used to reduce the cost of feed for dairy replacements.

The yield of beet tops is in the region of 20 to 25 tonnes per hectare. In certain instances these can be used to save the equivalent of 6 tonnes of silage, representing a financial saving in the region of £120 per hectare. This kind of saving makes the difference between sugar beet being justified or not justified on the farm as a whole. Note that £120 is equivalent to approximately 4 tonnes of sugar beet, which in turn is equivalent to approximately 11% of the average national yield per hectare.

CALF REARING

A consideration of the place of cattle other than dairy replacements on the farm starts with calf rearing. Calf rearing is competitive for labour and capital but not for land. The rearing of surplus calves to three to six months of age is a means of increasing the overall farm profit if labour and buildings are available. Calf prices fluctuate considerably from year to year as well as seasonally, and the ability to rear calves in a period of depressed calf prices lends stability to a dairy farm business.

Periods of prosperity in dairy farming are usually followed by periods of low profitability, often triggered by low calf prices, eg during the BSE scare of 1990. The prudent dairy farmer therefore always tries to be in a position to rear calves if and when this occurs, as the period of low prices is often short-lived.

Having reared calves the temptation is keep them too long. As a general rule, any calves born during the winter period should be sold before the end of the following July so as to obtain the high seasonal prices. This will be good from a cash flow as well as from a profit point of view, because it coincides with the time by which fertilisers should have been paid for, and on a tenanted farm this follows closely after the need to find cash to pay the rent bill.

It is very difficult to establish what contribution calf-rearing makes to overall farm profits because calf prices vary enormously from week to week. Calves are sold as 'calves' whether they are a few days old or two to three months of age. It is up to the individual manager to try to determine the price he can get for week-old calves in his locality, add the cost of rearing and then make a judgement as to whether the sale price is likely to justify the additional time and effort. The latter will certainly be needed but is not likely to be reflected in any significant increase in fixed costs; the main costs to consider are the variable costs of rearing the calves. An indication of what these are likely to be is given in Table 7.4.

Table 7.4 Main variable costs of calf-rearing

		Rearing to 3 months of age £ per calf		*Rearing to 8 months of age £ per calf*
Milk substitute	13 kg @ 115p	15	13 kg @ 115p	15
Other concentrates	150 kg @ 18p	27	150 kg @ 18p	27
Additional concentrates			250 kg @ 16p	40
Total concentrate feed		42		82
Opportunity cost of hay	60 kg @ 7p	4	260 kg @ 7p	18
		46		100
Vet fees and sundries		8		15
		54		115

BEEF CATTLE

Having reared the calf to eight months of age, we now have to decide whether it is worth keeping any longer. To make this decision we need to assess the margin we can expect from keeping the cattle and compare this to the opportunity cost of this decision to keep them. In short, we need to consider what else we could do with the land and capital that would be used to keep the cattle.

An eight-month-old calf should weigh in the region of 200 kg and could be worth £280 to £300 (£1.40 per kg) if sold in April to May. The costs incurred in keeping it for a further six months are relatively small—say £8 for grazing variable costs, £7 for concentrate feed, and £15 for interest on capital, but the price of beef cattle falls to a low in the autumn and the outcome could be as shown in Table 7.5, that is no extra profit.

Table 7.5 Beef cattle margins—summer

		£
Value at 12–14 months	300 kg @ £1.10	330.00
Value at 6–9 months	200 kg @ £1.50	300.00
Output	100 kg @ £0.30	30.00
LESS Feed	50 kg @ 14p	7.00
Grazing	0.1 hectare @ £80 per ha	8.00
Interest on capital	(£300* for half-year @ 10%)	15.00
		30.00
MARGIN TO COVER FEED COSTS		Nil

* To be absolutely correct we should also include interest on the cost of the food and grazing as well as the sale value of the animals.

The effective price received for the additional liveweight gain is only 30p per kg due to the fall in the overall price per kg. The extent of this fall is always difficult to predict as it varies considerably from year to year, according to trends in beef supplies relative to demand, and to the availability of winter feed supplies. The fall in most years is substantial and the general rule would be to sell the cattle before the autumn unless one can be sure of having adequate winter feed supplies to carry them through to the following spring. Note: Selling in mid-summer/early autumn also tends to be the best time to obtain the full value of the Beef Special Premium.

The output and costs incurred during the winter period depend upon whether the animal is kept on a store or finishing ration and, in turn, this depends to a large extent on whether they are in-wintered or out-wintered. The results could be as shown in Table 7.6 and as can be seen, the output is very dependent on the extent to which the value per kg changes between the autumn and the spring.

Table 7.6 Beef cattle margins—winter

	Store system		*Finishing system*	
Value at 18–20 months	400 kg @ £1.15 = 460		450 kg @ £1.15 = 517.50	
Value at 12–14 months	300 kg @ £1.10 = 330		300 kg @ £1.10 = 330.00	
	100 kg @ £1.30 = 130		150 kg @ £1.25 = 187.50	
LESS Variable costs:				
Feed	300 kg @ 14p	42	600 kg @ 14p	84
Sundries		8		8
Forage costs		*		*
Interest on capital*	£330 @ 10% for half-year	16.5	£330 @ 10% for half-year	16.5
		66.5		108.5
MARGIN TO COVER FIXED COSTS AND FORAGE COSTS		63.5		79.0

* See text for explanation.
** To be absolutely correct one should also include interest on the cost of the food and grazing as well as on the sale value of the animals on hand at the start of the costing period.

SHEEP

For many years now a breeding ewe enterprise has been very difficult to justify on the vast majority of dairy farms as it is very competitive for grazing in the spring, a critical time for dairy farming. They produce a low gross margin per hectare and cannot compete with cereals on an area basis. They tend to be less profitable than beef so are really only justified where there is an area of grassland that is not suitable for dairy cows and where there are inadequate buildings to keep dairy replacements and beef cattle.

Such an area of land has become available on some farms as a result of the EC reforms. There is now land on many dairy farms that is not eligible for cereal Area Payments and this makes breeding ewes more competitive relative to cereals. Stocking rate restrictions also make them more competitive relative to male beef cattle. The expansion/retention of an existing sheep enterprise at the present time is more readily justified on a dairy farm than in the past. Forward-thinking dairy farmers with land more suitable for sheep than dairy cows have therefore already purchased sheep quota or are making plans to do so.

The grazing of pastures by sheep in the autumn produces a considerable benefit but this coincides with a very busy time on dairy farms, particularly with autumn-calving herds. Fencing permitted, one of the best ways to gain this benefit to the pasture is to provide autumn and winter 'keep' facilities for sheep farmers on condition that they are responsible for shepherding. The author is privileged to see many farm management accounts: the sum received for 'sheep keep' is in itself quite modest relative to the milk cheque and the cattle sales *but* is often quite substantial relative to the 'bottom line', ie profit.

NON-LAND-USING ENTERPRISES

Many dairy farms are relatively small and have been made smaller by the introduction of quotas; consequently they are not large enough to provide a living for all the family without the introduction of a non-land-using enterprise.

Calf-rearing is such an enterprise and the possible returns from it have already been discussed. Intensive bull beef is another possibility. Its introduction/expansion will depend on whether the Beef Premium can be obtained either directly or indirectly, as previously discussed.

Pigs and poultry are possible enterprises. Which one is introduced will depend on the interests of the family or farmer concerned. Egg production is now very specialised and its introduction is unlikely. The introduction of a small table poultry or turkey enterprise concentrating on the Christmas market can, and does, offer a very useful contribution to profits. However, a pig enterprise can be very complementary to the dairy operations, particularly as a means of saving fertiliser costs. These costs are in the region of £50 to £70 per cow and most of these can be saved on an intensive pig and dairy farm by making effective use of the pig manure. On an all-grass farm there would also be potential benefits to be gained from sharing the machinery costs and/or organic irrigation costs of both cow and pig manure handling systems.

NON-FARM ENTERPRISES / DIVERSIFICATION

This chapter is concluded by a brief mention of the need to consider non-farm operations such as providing tourist facilities, whether it be bed and breakfast, a caravan site or other diversification enterprises such as horse livery. As with non-land enterprises, the objective here is to make a profit from resources which would otherwise be idle, such as an ex-farmworker's cottage or part of the large farmhouse, and/or redundant buildings.

An important point to remember when setting up these enterprises is that the income received from the customer will be subject to VAT. This enterprise should therefore be kept completely separate, from an accountancy point of view, from the main farm business, ie it should be set up in the wife's, son's or daughter's name so they can claim exemption from the need to charge VAT. (Note: The income that can be obtained from a business before registration for VAT is obligatory was increased from £37,600 to £45,000 in November 1993. If redundant farm buildings are to be used the repairs should be carried out in the first instance by the main business so that VAT on these expenses can be claimed, and the buildings should subsequently be let to the secondary business.)

Businesses such as these are now the ideal source of pin-money, and replace the traditional poultry enterprise of years ago. Whether or not these enterprises are successful will depend largely on the location of the farm and the interest and ability of the entrepreneur setting up the business. The world 'entrepreneur' has been used on purpose as skills such as marketing and getting on well with non-farming people are a pre-requisite to success. If these abilities are there then these enterprises can be exceedingly successful.

SUMMARY

To summarise, 'other enterprises' have become more important on the dairy farm since the introduction of milk quotas, and this importance has been further underlined by the EU reforms introduced in 1993.

Dairy farmers need to look increasingly at the spare resources that are available after the requirements of the dairy herd have been met. Successful use of these resources could be the key to success of the farm as a whole, particularly on the smaller farm.

CHAPTER 8

Farm Business Analysis and Planning Techniques and the Dairy Farmer

COMPARATIVE ACCOUNT ANALYSIS

A brief history of the development of comparative analysis techniques was given at the beginning of Chapter 3, the basis of comparative account analysis being 'What one farmer can do, another should be able to do too!'

There are several limitations to the value of comparative analysis as a method of improving farming profitability: the most important is that it is historical and in periods of inflation the information provided by these accounts is out of date before it is printed. Nonetheless, it provides a valuable analysis tool for the farm management specialist called in to advise a farmer on the future development of his business. By comparing the farmer's results to standards, an adviser can quickly identify weaknesses in the system and in enterprise efficiency. Comparative analysis, however, tends to tell only what is wrong with a business; it does not necessarily tell what needs to be done to put it right, nor does it always show why it is wrong.

Having identified the weaknesses and strengths in a business, the next step is to identify the reasons for these weaknesses and then take appropriate steps to put them right. These steps may necessitate a complete replanning of the business organisation and will involve budgeting and other management techniques.

A second problem with comparative account analysis is the wide range of performance concealed by the so-called average data. This is rectified to some extent by showing separate results for, say, the top 25% or bottom 25% of the sample. Often, however, it has to be accepted that there is no such thing as a typical average farmer.

Most financial survey results are presented in terms of results per hectare; farms are classified as being above average or below average according to their result per hectare, and this is defined as net farm income or management and investment income. An above-average level is then assumed to equate with 'above average' management ability and efficiency. This to a large extent is true but what is not properly taken into account is the considerable additional quota and capital invested per hectare that is associated with the good result. The additional capital invested and quota available are responsible for a significant proportion of the additional profit and there is need for a new definition of profit to take these into account. Alternatively, much more attention has to be paid to return per £100 capital or per litre.

The need to look carefully at capital investment per hectare is particularly true in dairy farming where high profits per hectare tend to be associated with higher stocking densities and therefore higher capital investments per hectare. Care in this connection has to be taken to compare farms of a similar size because in general the smaller the dairy farm the greater the intensity and the greater the profit per hectare.

The limitations of judging farms according to how much profit they make per hectare are even more pertinent following the introduction of milk quotas, as the key factor determining profit per hectare is now the amount of milk quota owned/leased per hectare. Despite this, one still receives glossy publications, including those from Genus, ranking farms according to profit per hectare with no mention of milk quotas, and dividing farms into size-groups based on the area farmed. What is wanted now are surveys in which the farm size is determined by the quantity of quota owned/leased, not the size in hectares.

To sum up, considerable care has to be taken when comparing your results to 'standards'. Remember that the best standards are the results you achieved on your farm last year. More is to be gained by studying in detail the differences in results of a few farms you know well than by studying highly technical and computerised results from a large number of farms following different systems and farming in different environments. Beware in particular dairy enterprise results quoted from data not backed up by full farm costings, particularly those dealing with MoC results.

One final point before leaving comparative analysis: the gross margin totals are usually fairly reliable and not dependent on many subjective judgements other than stock valuations. The fixed costs, on the other hand, are very dependent on subjective judgements and are much less reliable. Labour costs on small dairy farms are not assessed very accurately due to the high proportion of labour supplied by the farmer. Power and machinery costs too can sometimes be more dependent on the number of sons, and hence car and domestic fuel bills incurred, than on the actual costs of running the farm. Rent charges are also arbitrary on owner-occupied farms and are seldom worked out in relation to the actual quality of the land and buildings available. Finance charges are based on the actual sum that happens to be borrowed, not on the capital actually invested. They therefore tend to reflect how long the farm has been established, and/or recently expanded, not the efficiency with which the farm is managed.

Despite these reservations, some comparative standards are essential if you are to assess your management capability and competence as a dairy farm manager.

THE GROSS MARGIN SYSTEM AND ITS LIMITATIONS

The gross margin system, as such, had not been invented when the author attended college and taught farm management in the late 1950s and early 1960s. Its development in and since the 1960s has had much to do with the progress of farm business management. Virtually everyone in farming now talks about and uses gross margins and assesses enterprise performance in financial as well as physical and husbandry terms. Criticism is levied at, and in this book attention is drawn to, the limitations of the term 'margin over concentrates' but 30 years ago very few people had heard of it.

The main advantage of the gross margin analysis and costing system is its simplicity and the ease with which data can be collected and used for planning purposes. It is, however, much more suited to the needs of the arable farmer than to the needs of the dairy farmer.

The gross margin system is of most value when considering minor changes in the cropping and stocking of a farm, as for example whether to grow four hectares of barley instead of four hectares of wheat, or keep eight more cows. Such changes do not lead to any consequent changes in fixed costs, so the net effect on the profit is simply the change in the gross margin—but is it? In the case of the choice between wheat and barley there is no problem. The change is equal to the difference in the gross margin because such a change is not likely to affect any of the fixed costs, *but* in the case of the additional cows there would also be a consequent change in finance charges. This consequent change in finance charges is often forgotten because they are placed in fixed costs 'for convenience'. Finance charges *do* change in proportion to a small change in the size of an enterprise but the problem is that they *cannot* be easily allocated.

The choice between eight more cows or four more hectares of cereals should be budgeted, as shown in Table 8.1. A charge also has to be included against the cows for the cost of milk quota: either the actual cost of leasing, or the additional finance on milk quota that will need to be purchased. This together with the finance charges accounts for 50% of the potential gross margin from the dairy cows, leaving a net favourable margin compared to wheat of only £1,230.

The net favourable variance is very dependent on quota charges. If these fall to 3.6p per litre (the finance charges on 48p at 7.5%) the favourable variance increases by £1,150, that is 46,000 litres at 2.5 pence per litre. If on the other hand quota leasing costs rise by 2p to 8p there is an increase in leasing costs of £920 resulting in a net favourable variance for cows compared to wheat of only £310.

There is a tendency to assume that fixed costs, other than finance, will stay the same for *substantial* changes in the size of individual enterprises, eg an increase in cow numbers from 70 to 100. This may be true on a farm that is not using its buildings, labour *and* machinery to capacity, but on most farms it is only partly true. Most

Table 8.1 Comparing gross margins and finance charge to choose between enterprises

	£	£
Gross margin from 8 cows @ £925* per cow		7,400
LESS finance charges to fund 8 additional cows = £10,000 @ 10 per cent	1,000	
and quota leasing/finance charges 46,000 litres @ 6p	2,760	3,760
Net contribution from 8 additional cows (A)		3,640
Gross margin from 4 hectares wheat @ £620 per hectare		2,500
LESS finance charges to fund 4 hectares = £900 @ 10%		90
NET contribution from 4 hectares wheat (B)		2,410
Total A – B = increase in profit by keeping 8 cows instead of growing 4 hectares wheat		1,230

* Based on a yield of 5,750 litres (quota required 5,500 litres, threshold 4.5%).

farmers making such a change would have to incur some additional expenditure on buildings and equipment, with a consequent change in the mortgage and/or rent and an increase in machinery depreciation. The labour bill and other overhead costs would, however, probably not change.

Mistaken conclusions are often drawn when analysing accounts prepared on the gross margin/fixed cost basis. An example of this would be on a farm where the fixed costs were £1,000 per hectare and the gross margins were £1,800 for dairy cows and £650 for dairy followers. An understandable reaction would be that the dairy followers were not covering the allocation of fixed costs per hectare. In fact, the situation could be that the dairy followers were run on rough grazing land which was completely inaccessible and unsuitable for the dairy herd, and if not used by followers would have no alternative use. Discontinuing the youngstock enterprise would probably lead to *no* reduction in fixed costs other than finance charges and a loss in profit equal to the youngstock gross margin less finance charges.

Another category of stock which has suffered as a result of the gross margin system has been the sheep enterprise. The sheep gross margin is quite low and compares unfavourably to dairy cows and youngstock. Sheep can utilise marginal grazing unsuitable for the dairy herd and there is a bonus difficult to measure resulting from the use of sheep flock to take off all the surplus autumn grass and prevent the possibility of winter kill in short-term leys. Sheep will also help to establish maiden seeds prior to grazing by the dairy herd. In this connection, the reader is reminded that the gross margin system is simply a means of trying to implement some of the economic principles outlined in Chapter 2. Unfortunately, the system is so simple that most people forget the complementary and supplementary relationship between enterprises discussed in Chapter 2 and Chapter 7. They also forget that the fixed costs do change if one makes substantial changes in the cropping or increases the intensity of stocking. This may not happen in the short term but in the long term there is a tendency for fixed costs to catch up with the increased intensity, eg a bigger forage harvester is purchased, say, two years after expansion because it is justified by the expansion in herd numbers. Two years earlier one of the main reasons for justifying expansion was the opportunity it would present to 'spread machinery costs'!

PARTIAL BUDGETING

Prior to the introduction of the gross margin systems, the principle of fixed and variable costs was implemented by means of partial budgets. A partial budget is set out in Table 8.2, which illustrates the same example as shown in Table 8.1, that is the replacement of four hectares of wheat by eight dairy cows.

Table 8.2 Partial budget to assess effect of replacement of 4 hectares wheat by 8 dairy cows

Extra costs		£	*Costs saved*		£
Replacement heifers	2 @ £1,200	2,400	Wheat seed	4 @ £40	160
Concentrate feed	8 @ £250	2,000	Wheat fertilisers	4 @ £75	300
Veterinary and sundry costs	8 @ £75	600	Wheat sprays	4 @ £80	320
			Wheat sundries	4 @ £30	120
Forage costs	4 ha @ £150	600			
Labour costs		Nil	Labour costs		Nil
Power and machinery costs		Nil	Power and machinery costs		Nil
Finance charges	8 @ £1,250 @ 10%	1,000	Finance charges	4 @ £225 @ 10%	90
Milk quota leasing		2,760			
Revenue foregone:			Revenue gained:		
24 tonne wheat	@ £100	2,400	Milk sales	8 @ £1,300	10,400
Area payment	4 @ £250	1,000	Calf sales	8 @ £200	1,600
			Cull cows	2 @ £500	1,000
		12,760			
Net gain from change		1,230			
		13,990			13,990

The main advantage of the partial budget is that one considers each cost item in turn and asks 'In what way will this cost item change as a result of the change in cropping and stocking'? The main disadvantage is that it is a cumbersome and time-consuming way of considering several alternatives, eg barley compared with cows, compared with wheat, compared with dairy replacements. It is, however, by far the best way of assessing the effect of proposed changes in management strategy on the likely profit.

One of the problems with partial budgeting, as with all planning, is the need to make decisions without really having the knowledge or time to work out the effect of the proposed change on the profit. For example, you should pose the question, 'Will it pay me

Extra costs	*Costs saved*
Machinery depreciation Interest on additional capital invested in machinery Fuel and repairs Labour	Contract charges Concentrates
Revenue foregone	Revenue gained Milk

to purchase my own equipment to make silage rather than use a contractor'? A partial budget setting out the headings to help answer this question is given above. The extra costs would be those associated with the purchase and use of the silage-making machinery, including machinery depreciation, fuel and repairs, and the additional labour costs that might be incurred. Against these one can offset the savings in contract charges and the expected benefits 'better silage' would have on the milk sales and concentrate costs. The latter are the most difficult items to evaluate. Having worked this out, one would then have to consider the question 'Would it pay better to invest the capital in more cows and/or another enterprise'? etc, etc.

In practice, there is not enough time to evaluate all the alternatives. Computers can be used to calculate 'What if' situations. The danger then is, that the issues become too confusing. 'Back of an envelope' calculations, despite modern technology, are still often the best way to make decisions, particularly if one has to back one's hunches as to possible future trends in profitability. For example, the partial budget in Table 8.2 shows only a modest advantage for cows compared to cereals, as it is based on expected 1994 prices. In 1992 the cows could have been purchased for £800 to £850 each, £400 less than in 1994, and the quota could have been purchased for 30p to 35p per litre @ 8% equals a cost per litre of only 2.4p to 2.8p, not 6p. Making a decision to expand cow numbers in 1992 was much easier than in 1994, albeit that profits from dairy cows at that time were much lower than in 1992. Budgeting is only part of the decision-making process; getting the timing right is of equal if not more importance.

COMPLETE ENTERPRISE COSTINGS

Prior to the introduction of the gross margin system it was normal practice to carry out complete enterprise costings which involved allocation of fixed costs to individual enterprises as well as variable costs.

The time taken to carry out full enterprise costings is much longer than for gross margins and much of the information produced is of little value. However, on the larger dairy/mixed cropping and dairy farm there is a growing need for 'fixed costs standards' for individual enterprises as well as fixed costs standards for farming systems.

Farmers are exhorted to watch their fixed costs but there is very little information available to indicate what the fixed costs should be at the individual enterprise level. Most herd managers, for example, can readily quote their MoC per cow and per litre, stocking rates, calving index, etc, all figures related to the gross margin. Very few know the cost of labour per litre, or their power and machinery costs per litre, which are of equal, if not more importance. One of the reasons is the time-consuming nature of the records that are required, but this is one way in which computers may come to be used more extensively in the future, ie to investigate costs that cannot be determined at the present time due to the expense of recording this information. It is stressed, however, that this is only likely to be necessary on the larger and the mixed dairy/arable farm.

A second reason why farmers and managers place too much emphasis on marginal data is because there are a multitude of advisers working for feed and fertiliser firms who are prepared to work out the MoC per cow, per litre or per hectare to promote the sales of their firm's products, and often these advisers are not aware of the whole picture. This is also true of many independent advisers whose work is solely at the enterprise level, and concerned mainly with technical efficiency.

There is still, for example, much too much emphasis on such items as 'yield from forage'. Commentators, for example, on recent costing carried out by national organisations have lamented the drop in yield from forage even though the yield per cow and the margin per cow have improved. It needs to be remembered that the financial objective is to make money from cows, not to produce as much milk as possible from forage.

SEPARATE COST CENTRES

This was mentioned briefly in Chapter 1 when discussing the management structure of the larger farm. The setting up of separate bank accounts for each of the major enterprises is strongly recommended on the larger farm, as a much better method than complete costings of identifying strengths and weaknesses in the farm business as a whole.

CASH FLOW AND MANAGEMENT BY OBJECTIVES

In Chapter 1 the two basic functions of management, namely deciding what to do and doing it, were discussed.

The efficient dairy herd or dairy farm manager does not simply compare his results to standards. He prepares a physical and financial budget target for the year, month, or week and then compares his result to the target. This applies whether he is preparing the budget for the whole farm or comparing his level of intended feeding of the cows to what actually happened.

Experiences in the field of consultancy and managing one's own business have shown that this is by far the most important aid to successful dairy business management. The important thing is to plan what to do and make it happen. How this is done is described in detail in the second half of the book.

From a cash flow point of view, what one does is write down the actual cash flow for the previous year in one column and the budget objective for the year to come in another. The art of management is then to see that these budget results happen. In this connection, it is most important to distinguish between a budget and a forecast. The latter simply shows what will happen if no changes are made in organisation or business methods, the former shows the need for, and leads to, changes in how the business is organised and managed.

Reference was made in Chapter 3 to costings produced by Manchester University and agricultural economic departments of other universities. Reports are also published from time to time by consultancy organisations which show the trends in their clients' financial results over a period of time. These simply appear to follow the national trend in costs and returns and do not appear, to a large extent, to demonstrate the effect of the consultancy organisations on the results achieved by their clients. In other words, the costings and forecasts produced by these organisations appear to be an end in themselves and not a means of making things happen.

THINK CASH FLOW

To think cash flow is probably the most important lesson the author has learned as a businessman as opposed to an academic. The availability of cash determines whether or not there is money to pay the wages next week, or to buy cows to see that the target MoC is achieved.

To date, the book has essentially been about items of academic and general interest such as complementary and supplementary enterprises. Discussion has taken place in some depth as to the value of partial budgets compared to gross margin analysis.

Whether margin over concentrates per cow or margin per acre is the better method of measuring efficiency has been discussed at length, as all these are matters that are of great interest to many academics and other advisers working in the field.

The job of management, however, is to go out, set up a business, prepare the cash flows and make a profit. This is what the rest of the book is about.

SECTION TWO

Making Profits by Putting Principles into Practice

CHAPTER 9

Taking Over a New Farm

ASSESSING THE FARM AND ITS RESOURCES

In this chapter it is assumed you are taking over the management of an 84 hectare dairy farm. The objective is to deal with the main points you should consider when arriving at your management strategy. The general approach would be much the same if you were taking over as a tenant, or were a son/daughter looking at ways to make the home farm capable of providing a living for you as well as your parents.

When taking over a farm you need to guard against imposing a preconceived plan on it. It is most important that you assess all the farm resources carefully and then prepare a plan to suit, with one exception—milk quota.

(a) Milk quota

In the short term, plans for the farm may need to be based on the quota available, but in the long term it is most important that this should be made to fit the farm, and not vice versa.

There will be years when quota leasing charges seem high relative to dairy cow profits, as was the case in 1990/91 and 1991/92, but farming through these difficult years with enough cows to fit the farm allows good years like 1992/93 and 1993/94 to be exploited. If the buildings and other resources are appropriate to 120 cows but there is quota for only 100, the aim should be to keep 120.

(b) The land

Examine the nature of the land and try to assess the inherent fertility and the suitability of each field for grazing, conservation or other crops. In particular assess whether there is a substantial area that cannot be grazed by the dairy herd; this will indicate whether there is a basic need for a youngstock enterprise. You should also assess the difference between the total *effective* hectares and the total farm hectares including roads, buildings and woodlands and waste areas. The effective area may be 5 to 10% less than the total area and this needs to be taken into account in arriving at the rent that can be afforded.

Check which fields are eligible for Arable Area Payments; this will be particularly significant if these fields are not suitable for cow grazing and/or silage. Summarise this information as shown in Table 9.1.

In addition, you need to take careful note of which fields require drainage and/or other improvements. Drainage and pasture quality may limit cow numbers in the short term. It is important to try to assess whether poor swards are due to poor management or poor inherent fertility.

Table 9.1 Assessing the land

		Suitable for:				
Field no. and general remarks	*Area hectares*	*Cow grazing but not conservation*	*Cow grazing or conservation*	*Youngstock grazing or conservation*	*Youngstock grazing only*	*Cereals*
1 Badly poached	6	6	—	—	—	—
2 Good ley	8	8	—	—	—	—
3 Needs drainage	4	4	—	—	—	—
4 Poor water supply	8	—	8	—	—	—
5 Good ley	6	—	6	—	—	—
6 Good ley	4	—	4	—	—	—
7 Poor ley	8	—	8	—	—	8
8 Not eligible for AA† payments	8	—	8	—	—	8
9 Poor access	8	—	—	8	—	8
10 Poor access	4	—	—	4	—	4
11 Poor access	6	—	—	6	—	6
12 Unproductive	8	—	—	—	8	—
13 Unproductive	2	—	—	—	2	—
	80*	18	34	18	10	34

* Plus 4 hectares buildings, roads and waste areas
† Arable area

(c) Water supplies, roads and fences

These must also be taken into account when assessing the current capability of the land and its future potential. A map of the farm, scale of 1 in 10,000, is an invaluable aid in preparing the cropping plan.

(d) Buildings and fixed equipment

Assessment of these needs to deal with the following:

1. Present use, potential use with minor capital improvements, potential use with major capital improvements.
2. Number of cows, youngstock and calves which can be accommodated in the buildings.
3. Ease or difficulty of feeding and managing the stock.
4. Amount of fodder that can be stored in the buildings as silage, hay, straw or grain.

You should make a sketch plan of all the buildings on which materials and cow-flow diagrams can be superimposed as part of the assessment.

(e) Farm cottage(s)

The availability or otherwise of these will influence the choice of staff.

As a result of the above resource assessment you should be able to formulate a short-term cropping and stocking programme appropriate to the existing farm resources. From this a longer-term cropping and stocking programme can be derived, given (1) minor capital and husbandry improvements, and (2) major capital and husbandry improvements.

(f) Farm staff

On a farm of this size there will be one or two staff already employed on a full-time or regular part-time basis. Whether you should continue with these staff may be a major policy decision. If you are appointed as manager it is likely that your employer will wish you to make effective use of the men already on the farm. This may well necessitate revising the policy you prepared based on your initial assessment of the farm.

(g) Machinery and equipment

You should prepare an inventory of the machinery and equipment already on the farm. Note its state of repair, and assess its value and its suitability to the farming system you have in mind.

Estimate the capital cost of any additional machines that need to be purchased together with the cost of those in need of replacement. The depreciation charges resulting from these proposed transactions can be calculated along the lines of the example in Table 9.2 but the most important part is the proposed cash outlay. At £11,000 this happens to be closely in line with the depreciation charge; in practice it could be much higher.

Table 9.2 Machinery inventory

Machine	*Notes*	*Present value (A)* £	*Proposed net outlay (B)* £	*Total (A+B)* £	*Budget depreciation* £	*Budget closing valuation* £
Tractor		8,000		8,000	2,000 (25%)	6,000
Tractor	Needs replacing	4,000	6,000	10,000	2,500 (25%)	7,500
Tractor		2,000		2,000	500 (25%)	1,500
Forage Harvester	Poor maintenance	3,000		3,000	600 (20%)	2,400
Silage trailers (2)		3,000		3,000	600 (20%)	2,400
Mower	Needs replacing	800	1,600	2,400	480 (20%)	1,920
Heavy roller		1,000		1,000	200 (20%)	800
Fore-end loader		1,200		1,200	240 (20%)	960
Fertiliser spreader	Scrap value	Nil	1,600	1,600	320 (20%)	1,280
Plough		1,000		1,000	200 (20%)	800
Cultivating equipment		1,000		1,000	200 (20%)	800
Sprayer	Scrap value	Nil				
Slurry tanker		1,000		1,000	200 (20%)	800
FYM spreaders (2)	Poor condition	2,000		2,000	400 (20%)	1,600
Farm vehicle	Poor tyres	3,000		3,000	750 (25%)	2,250
Haymaking equipment		Nil				
Baler		—		—	— —	—
Milking equipment		8,000		8,000	800 (10%)	7,200
Bulk tank		3,000		3,000	300 (10%)	2,700
Contingency			1,800	1,800	360 (20%)	1,440
TOTAL		42,000	11,000	53,000	10,650	42,350
Per hectare*		530	137.57		134	

* Effective area excluding roads and buildings.

If you are taking over the farm as a tenant the inventory will consist of machines you already own, and could be nil. Your capital will be limited and the proposed investments will need careful scrutiny to ensure that adequate capital remains for more important investments in livestock.

(h) Livestock

The procedure you should follow to assess your stock is as follows:

1. Count how many there are and judge the condition of each one.
2. Make notes on their production potential, bearing in mind that you may not have any historical data. This could well be the first real test of your ability as a stockman to pick out the good cows from the poor ones without having detailed records to do the job for you.
3. Prepare a simple inventory and establish the value of the stock.
4. Compare this inventory and valuation with that required for your policy.

(i) Crop produce and stores on hand / tenant right

The quantities on hand will depend on the time of year.

As the prospective manager you will be mainly interested in determining whether or not food supplies are adequate and in checking whether seeds and fertilisers for the coming year's crop are on hand or not.

If you are a prospective tenant this area of assessment is vitally important because it will have a considerable effect on the capital required to take over the farm.

As an incoming tenant you will be expected to pay the market value for all crop produce and stores on hand. In addition payments may well be necessary in respect of unused manurial values and tenant's fixtures. This last point takes us back to the assessment made of the water supplies, fences, hedges, buildings and other fixed equipment. As a prospective tenant you will have a keen interest in determining the likely capital sum you will need to pay the outgoing tenant for his 'tenant's improvements and fixtures'. In addition you will have made enquiries and taken note of any dilapidations that may be charged by the landlord or that you may be taking over from the existing tenant.

You will also need to look again at the buildings as the landlord may be prepared to carry out improvements provided you are also prepared to pay the appropriate increase in rent to fund these improvements.

ANALYSIS OF PREVIOUS YEAR'S FINANCIAL PERFORMANCE

Taking over as manager you should have access to the previous year's financial results. These need to be examined with these particular points in mind:

(a) What are the likely fixed costs? This information is going to be vital in the next stage of the venture—preparing your first year's programme together with the cash flow/trading budget.
(b) What performance level can you expect from the individual livestock enterprises? Are there major weaknesses in enterprise management? If so can they be corrected quickly, or do they necessitate below-average results in the future until such time as the inherent problems leading to them can be overcome?
(c) Are there basic weaknesses in the farming system such as too many youngstock or too few cows? If so, is this simply due to a poor understanding of the profitability factors in dairy farming? Or is it due to lack of resources in the form of working capital to purchase the stock or fixed capital needed to provide additional buildings and fixed equipment?
(d) Finally, and most important, is it because in the past, farming has been made to fit the quota?

The basic objective of the analysis is to gather information that will allow you to do the following:

1. Prepare a physical and financial budget for your first year of operation that is likely to be fairly accurate.
2. Quickly identify weak areas in enterprise management that can be corrected before they occur again.
3. Decide on a farming programme that is appropriate to the needs of the business.

ASSESSMENT OF CAPITAL POSITION

It is essential to make this assessment *before* moving on to the detailed planning of the cropping and stocking of the farm, and the preparation of the whole farm budget for this detailed plan. If not, you could find you have prepared a detailed plan that is completely outside your own financial resources as a prospective tenant, or the financial resources of your employer if you are a manager.

In the case of a farm manager or a son/daughter, the first stage in this assessment is to detail and evaluate the farming assets already on the farm, and at the same time to list the liabilities in order to arrive at the net capital (net worth): see Table 9.3.

The second stage is to consider ways in which the allocation of

Table 9.3 Assets and liabilities statement for 80 hectare farm at 31 December 1993

	No.	*Per head*	*£*	*Total £*
Assets:				
Dairy cows	129	700	90,300	
Bulls	1	700	700	91,000
In-calf heifers	30	700	21,000	
Bulling heifers	10	550	5,500	
Yearling heifers	10	400	4,000	
Heifers 6–12 months	20	300	6,000	
Heifers under 6 months	10	200	2,000	
Beef cattle	Nil		—	38,500
	210			
Purchased feed			2,000	
Seeds, sprays and fertilisers			4,000	6,000
Home-grown hay 10 tonne			600	
Home-grown silage 100 tonne			1,400	2,000
				137,500
Debtors (previous month's milk cheque)				
Machinery and equipment (see Table 9.2)				12,000
Tenant's improvements (written-down value)				42,000
Milk quota purchased				15,000
Land owned (20 ha at cost 1980)				Nil
				100,000
				306,500
Liabilities:				
Loan for land purchase				40,000
Bank overdraft				40,000
Creditors				12,000
				92,000
NET WORTH (ASSETS)				214,500

capital resources could be changed to bring about a better liquidity position and/or potentially more profitable farming system.

The third stage in this capital assessment is the production of an annual cash-flow budget to ascertain how the asset requirement/ borrowed capital requirement changes during the year.

The approach of a prospective tenant farmer is very similar. In this case he lists the resources he proposed to employ on the farm to find the total initial capital requirement. In this case we must not forget the capital required to meet the tenant's ingoing.

The final stage usually involves discussing the budget with the bank manager to get his approval/agreement to the facilities required to implement the plan.

RECURRING NATURE OF RESOURCE ASSESSMENT

This chapter has emphasised the need for resource assessment when taking over a new farm. However, resource assessment is not simply a one-off study; it is a policy that should continue throughout your period of management. As time progresses the assessment becomes more detailed and more accurate because it is based on experience.

This assessment tends to become too inward-looking and often places too much emphasis on maximising profits in relation to the assets already available, particularly in relation to land and quota. Emphasis has already been placed on the necessity of not basing the future on the historical allocation of quota. The same is true, although to a lesser extent, in relation to land. There is a tendency to spend more and more capital and effort trying to increase the profit from the land already available, a tendency reinforced by the fact that most research and economic reports work out profits in relation to land as this is the most scarce resource.

A successful dairy business, however, soon reaches a stage when it needs more land as well as quota in order to expand. Time should therefore be spent seeking potential areas of land, or buildings other farmers are prepared to let on a short-term basis, the objective in the long term being to exploit the opportunity when it occurs of either buying or renting more land on a permanent basis.

RESOURCE ASSESSMENT SUMMARY

We need to summarise the resources available on our example farm, as itemised below:

1. *Milk quota:* 680,000 litres. The amount awarded to the farm based on the production in 1983 which was in the region of 840,000 litres, ie about 20% more than the current quota.
2. *Land:*

   (a) Of the 80 effective hectares, 24 are in permanent pasture. These are not suitable for conservation and 16 of these 24 hectares are inaccessible and therefore difficult to graze with dairy cows. The remaining 56 hectares can be ploughed and are well suited to intensive grassland production.

   (b) Of the 38 hectares suitable for cereals, 8 are not eligible for Arable Aid Payments. These 8 hectares are convenient for cow grazing so this is not a serious disadvantage.
   Forage maize could be grown but has not been grown in the past and a separate silage clamp would be required.

   (c) Only 40 hectares are conveniently placed for grazing by the dairy cows and this tends to make grassland management rather difficult.

3. *Buildings and fixed equipment:* Cubicle housing is available for 140 cows and there is an open, that is, not covered, grass silo suitable for self-feeding. Supplementary feeding outside the parlour is difficult. Out-of-parlour feeders have not been installed.
   The slurry storage facilities are not ideal and the milking parlour is becoming outdated. Additional capital expenditure is desirable to improve the parlour and to improve slurry storage.
   Youngstock accommodation is available for 70 youngstock. Special calf-rearing facilities are limited but use can be made of the cubicles in the summer period.
4. *Farm staff:* One man is employed besides the manager/farmer. The employment of a third person is dependent on being able to prepare plans and budgets to make this worthwhile, and at the same time have adequate funds to finance the desirable improvements to the parlour and slurry system.
5. *Machinery and equipment:* This is adequate but not substantial. Silage is made on a contract basis and this has worked well in previous years.
6. *Livestock:* Numbers on hand (see Table 9.3) are as follows:

Dairy cows	129
Beef bull	1
Heifers over one year	50
Heifers under one year	30

The cows are mainly Holstein Fresians and include a few that have simply been maintained to keep cow numbers up and really need culling. The young stock are mainly by Genus bulls but do not include a significant number of high index animals.

7. *Capital:* Repayment of the loan taken out to fund land purchase in 1980 has been more difficult than expected as interest rates have been high and it was purchased just before quotas were introduced. The bank would like to see this reduced fairly quickly now that interest rates have fallen. The interest rate is now 2.5% over base, ie a total of 8% (1994) compared to a rate being charged two years ago of 14–15%. The bank is reluctant to increase the overdraft as this too has tended to go up rather than down in the past.
8. *Previous year's performance:* Fortunately, up to date accounts for the past two years are available and these are summarised in Table 10.1 (Chapter 10).

There has been a very significant improvement in the profitability of dairy farming in the current year, 1993/94, due to both an increase in the milk price and improvements in the sale value of cull cows and calves. This is expected to be reflected in the results for this farm and the budget results for 1993/94 are summarised in Table 10.1 alongside the actual results for the previous two years.

For simplicity's sake, the stocktaking valuation and the number of cattle/litres sold has been assumed to be the same for all three years. In practice, this would not happen but keeping them the same helps to show the effect on profitability of the recent changes in cost/prices.

The question now is, can advantage be taken of this upturn in profitability to improve the long-term prospects of the farm? This is the subject of the next chapter.

CHAPTER 10

Preparing the Annual Budget and Determining Management Objectives

STRATEGIC OBJECTIVE

Suppose that another member of the family now wishes to join the family business and obtain his living from the farm. The farmer's existing employee, however, has been on the farm for nearly 20 years, is in his late 40s and is a good loyal worker. He is *not* to be made redundant. Plans have to be prepared that will allow three workers to make a living from the farm.

ASSESSMENT OF PREVIOUS YEAR'S PERFORMANCE

The farming system has in effect been made to fit the quota (see Table 10.1). Dairy cow numbers are 10 to 15 below the farm's building capacity and the stocking rate at 2.12 livestock units per forage hectare is less than that considered feasible. The yield per cow is also modest at 5,386 litres per cow and could be readily increased to 6,000 litres if a more liberal feeding regime was adopted.

The current year's profit is expected to be nearly £50,000 compared to a profit two years earlier of only £17,000. Note: It was low in 1991/92 in part owing to the effect of BSE on cull cow and calf prices. The tremendous improvement in profitability in 1993/94 reflects the very favourable trends that have occurred in prices relative to costs. Milk prices, cull cow prices and calf prices are all up compared to the previous two years and these increases are

Table 10.1 Summary of financial performance

			Estimate year ended March 1994		*Actual year ended March 1993*		*Actual year ended March 1992*
Average no. cows			130		130		130
Average no. youngstock			80		80		80
No. livestock units			170		170		170
and stocking rate (L.U./Ha)			2.12		2.12		2.12
		Unit price	£	*Unit price*	£	*Unit price*	£
Milk sales 700,000		21.5p	150,500	19.5	136,500	18.5	129,500
Cull cow sales	28	550	15,400	450	12,600	400	11,200
Surplus cow sales	2	1000	2,000	700	1,400	600	1,200
Calf sales	90	155.5	14,000	100	9,000	80	7,200
Other income			Nil		Nil		Nil
			181,900		159,500		149,100
LESS							
Livestock purchases			Nil		Nil		Nil
Dairy cow feed 170 tonne		150	25,500	150	25,500	150	25,500
Younstock feed 80 head		87.5	7,000	87.5	7,000	87.5	7,000
Vet, med and sundries							
130 Cows		80	10,400	80	10,400	80	10,400
80 Youngstock		25	2,000	25	2,000	25	2,000
Forage costs 80 hectares		165	13,200	165	13,200	165	13,200
			58,100		58,100		58,100
GROSS MARGIN			123,800		101,400		91,000
LESS							
Paid labour 1 man + casual			18,500		18,000		17,000
Power and machinery							
(inc contract) 80 hectares			18,500		18,000		17,000
Depreciation			12,000		10,000		9,500
Rent 60 hectares			8,000		8,000		8,000
Property repairs/improvements			3,000		3,000		3,000
Sundry overheads			8,000		8,000		7,500
Finance charges £80,000		7.5%	6,000	12.5%	10,000	15%	12,000
			74,000		75,000		74,000
PROFIT			49,800		26,400		17,000
LESS							
Private drawings			15,000		16,000		14,000
and tax provision			15,000		4,000		3,000
			30,000		20,000		17,000
PROFIT FOR REINVESTMENT			19,800		6,400		Nil

substantial but may prove to be temporary and future plans have to bear in mind that prices could fall again.

The improvement in profitability also reflects changes in interest rates, down from 15% in 1992 to 7.5% in 1994. Hopefully, these will continue to be low reflecting the government's determination to control inflation. The margin available for reinvestment/reduction in borrowing in 1994 is expected to be nearly £20,000 but the margin in 1993 was only £6,400, and in 1991/92 it was nil.

The depreciation charge included in the accounts for 1993 is £10,000 but this is based on historic costs. There is a need for investment during the next 3 to 4 years in the region of £20,000 to £25,000 per annum if the equipment and buildings are to be kept up to date, ie double the depreciation figure included in the accounts. This includes investment in an improved slurry system which will not make any contribution to future profits.

MISSED OPPORTUNITIES

1. When quotas were first introduced a decision was taken to rear heifers surplus to requirements. This strategy has been discontinued as the price received for surplus animals in the past did not seem to warrant the time and money involved. It is a different matter now, however, with prices for surplus cows/heifers in the region of £1,000 to £1,200 compared to only £600 to £700 two or three years ago. Even at £600 to £700 per head these heifers were making a contribution to the profit, probably in the region of £300 per heifer reared, which for ten heifers would have given an increase in the profit of £3,000 more per annum.
2. The extra heifers could have been kept to produce more milk, say, an extra 100,000 litres per year. This would have necessitated leasing quota but after taking this into account the increase in profit would probably have been in the region of 6p per litre, or an extra £6,000 per year. The gross margin achieved per litre would have been in the region of 15p less 6p leasing charges, which equals say 9p per litre, less increase in fixed costs of say 3p per litre, which gives an increase in the net margin of 6p per litre.
3. No milk quota has been purchased. Milk quota could have been purchased for no more than 30p per litre in 1993, compared to a current price of 45p.
4. Grants towards improvements to the slurry system of 50% could have been obtained but these have now been reduced to 25%.

Father and son are aware of these missed opportunities and do not want to miss any more. They have therefore called in the services of a farm business consultant to discuss and help formulate their future strategy. Father and son, although they get on well, have differing ideas for the future, father being generally the more conservative. The consultant has a role to play as an arbitrator as well as advisor.

LONG-TERM OBJECTIVES

At the first meeting held in November 1993, it becomes apparent that the son has ambitious ideas for the future. It was pointed out that these will take some time to put into practice but it is agreed they should be recorded to be borne in mind when considering short-term plans. The consultant is keen to do this as he is in broad agreement with the ideas put forward, which are summarised below:

1. As already mentioned, the overall objective is to provide a good living for three families.
2. To produce one million litres by the year 2000, preferably from not more than 150 cows, ie with a yield in excess of 6,500 litres per cow.
3. To improve the genetic potential of the herd to help achieve this objective.
4. To exploit the availability of cheap purchased feeds, particularly bulk feeds.
5. To expand the business in a tax-effective way.
6. To update the buildings and equipment for the above strategy and to be in a position to purchase and/or rent more land if and when this becomes available.
7. The son expects to marry in 18 to 24 month's time and his future wife has expressed an interest in making use of some of the buildings for non-farm activities. This is to be borne in mind when modifying the existing buildings.
8. Most important, to start and plan now for father's retirement in say 15 years' time. Funds have to be found to invest outside the business to provide a pension fund, and purchase a house leaving the son on the home farm as tenant.

BUDGETS AND FARMING PROGRAMME FOR 1994

The tax year is the end of March but it is agreed that in the first instance budgets should be prepared for the calendar year 1994. The actual results will be monitored against this budget and a decision taken at a later date whether to change the budget year-end to March. These budgets are to be completed ready for discussion before the end of January. The consultant is keen to do this as there are several decisions that should be taken before March 1994, from both a future profit and a tax-saving point of view.

A cash flow budget was prepared for 1994 and this is shown in Table 10.2.

Possible trends in prices/costs that were borne in mind in preparing the budget and recommendations for 1994 include the following:

1. Milk production in the 1994/95 quota year could be well above quota in the period April to October, leading to high quota leasing prices at the end of the year, and a possible reduction in dairy cow values. It is planned to produce above quota in 1994/95 so quota will be leased early in the season.
2. Less emphasis is to be placed by Milk Marque on high summer milk prices. (A decision has already been taken to join Milk Marque.) Less emphasis will therefore be placed on summer calving. It is planned to purchase 15 cows *before the end of March* to lift production above that achieved in the previous year. It is also planned to cull some of the cows with low genetic potential and replace these in the autumn with cows of higher genetic potential *if* the anticipated fall in prices occurs.
3. Dairy cow prices at the present time are at a record high so the first idea was to delay any expansion until the autumn. This, however, would mean that milk production could not be increased significantly in the quota year 1994/95, compared to 1993/94. It would also mean that another opportunity would be missed. Milk production on this farm for the 1993/94 quota year is 'just under quota'. It is agreed that a 'risk' should be taken to produce 2 to 3% above quota, ie 20,000 litres, which is the amount the additional cows purchased before the end of the year will produce. Timing of purchases will be arranged to achieve this objective, ie a production 2.5% more than quota.
4. It was also noted after discussion with the accountant that the purchase of these cows before the end of March will probably be tax effective. This will be particularly true regarding those which

TABLE 10.2 Cash flow budget of an example farm year ending December 1994

Details/date		*A Actual Previous Year*	*B Jan to March*	*C April to June*	*D July to Sept*	*E Oct to Dec*	*F Total*
Milk sales	1	150,500	40,000	40,000	46,000	46,000	172,000
Less concentrates	2	25,500	10,000	8,000	6,000	8,000	32,000
MoC	3	125,000	30,000	32,000	40,000	38,000	140,000
Cows	4	11,000	5,000	5,000	5,000	?	15,000
Calves and stores	5	12,000	3,000	1,500	4,500	3,000	12,000
Grain and Area Payments	6	Nil				6,000	6,000
Other receipt	7	Nil	–1,000	–1,000	–1,500	–1,500	–5,000
Total trading receipts	8	148,000	37,000	37,500	48,000	45,500	168,000
Capital introduced	9	Nil					Nil
Machinery sales	10	5,000					—
VAT net payments	11	1,000					—
Total receipts	12	154,000	37,000	37,500	48,000	45,500	168,000
Livestock purchases	13	Nil	20,000			?	20,000
Youngstock feed	14	7,000	2,000	1,500	1,500	2,000	7,000
Vet, AI, sundries	15	12,400	3,400	3,400	3,400	3,400	13,600
Forage costs	16	13,200		7,000	5,300	5,000	17,300
Paid labour	17	18,500	7,000	7,000	7,000	7,000	28,000
Power/machinery costs	18	18,500	5,200	5,200	5,200	5,300	20,900
Rent/property repairs	19	11,000	5,000	1,000	5,000	1,000	12,000
Sundry overheads	20	8,000	2,100	2,100	2,100	2,100	8,400
Milk quota leasing	21	—		3,000	3,000		6,000
Finance charges	22	7,000	2,000	2,000	1,400	1,400	6,800
Total trading payments	23	95,600	46,700	32,200	33,900	27,200	140,000
Private drawings/tax	24	20,000	6,500	3,500	6,500	3,500	20,000
Capital expenditure	25	17,000	3,000	—	—	?	3,000
Bank loan payments	26	1,400	600	600	600	600	2,400
(B) Total payments	27	134,000	56,800	36,300	41,000	31,300	165,400
(C) Surplus/Deficit	28	+20,000	–19,800	+1,200	+7,000	+14,200	+2,600
Op. Bank Bal. (Overdraft)	29	–60,000	–40,000				
Cl. Bank Bal. (Overdraft)	30	–40,000	–59,800	–58,600	–51,600	–37,400	

Notes

1	800,000 litres at 21.5 p. Same price as last year. Could be more.	1
2	800,000 litres at 4.0p	2
3	800,000 litres at 17.50p	3
4	30 at £500, lower price than that attained in 1993.	4
5	80 at £150	5
6	50 tonnes grain £4,600 plus £1,400 Area Payment.	6
7	Contingency MoC shortfall	7
8		8
9		9
10	Budget shown net at Line 23	10
11	Receipts equals payments	11
12		12
13	Budget 15 cows £18,000 (target 12 for £14,400) plus 1 bull £2,000 (net sale)	13
14	Same as previous year	14
15	£1,200 added for 15 extra cows	15
16	Same as previous year plus £2,500 bulk food and £1,600 for growing cereals	16
17	Include son's wages £8,500 + £1,000 for inflation in costs	17
18	Extra £1,500 for inflation and £1,400 for 8 ha cereals	18
19	Repairs up £1,000	19
20		20
21	100,000 litres at 6p	21
22	Loan interest £2,800, overdraft interest £3,400, charges £600	22
23		23
24	£14,000 drawings plus £6,000 tax paid Jan (£3,000) and July (£3,000)	24
25	Investment in Oct/Dec quarter will depend on progress	25
26	Repayments increased to £200 per month	26

are to be purchased to replace cows that are now to be culled because they are more or less at the end of their useful life.

It is anticipated that the purchase price of the cows will be in the region of £1,200 and some of these will be written down in the books to £500, giving a tax loss of £700 and a potential saving in tax of £175 per cow purchased (£700 at 25%).

BUDGET ASSUMPTIONS

The assumptions made in the farming programme/budget for 1994 compared to 1993 are detailed in Table 10.2 and are summarised below:

1. A modest increase is assumed in all fixed costs although a determined effort will be made to keep them at the same level as in 1993.
2. Capital expenditure on machinery and equipment is to be kept as low as possible. Some expenditure will take place before the end of the 1993/94 tax year but no more is planned before December 1994. A decision whether to go ahead and update the parlour will be made in the light of progress and the budget for 1995.
3. A target has been set to increase the MoC from £125,000 (700,000 litres at 17.84p) at the end of December 1993, to £140,000 (800,000 litres at 17.50p) by December 1994, but a contingency of £5,000 is allowed for in the cash flow in case this is not met (see line 7, Table 10.2). It could be difficult to achieve this by the end of December, but it is expected that it will be attained by the end of the quota year. An increase in milk price is possible in 1994/95 but it is considered prudent not to include this in the budget (see line 1, Table 10.2).

   To ensure the MoC target is achieved, the budget allows for three more cows to be purchased than may be necessary (see line 13, Table 10.2).

   A rule-of-thumb target has been set to produce 80,000 litres in May, or 2,580 per day, ie 10% of the budget set for the whole year.

   It is also agreed that in due course a detailed yield predicted should be prepared to determine just how many cows need to be purchased. (Note: The production of these herd yield predictions has been a common part of business management practice since milk quotas were introduced.) Appendix 3 shows how these can be prepared and used in practice.

   No mention has yet been made of budget yields per cow. This

is deliberate. The budget objective is to produce 800,000 litres. Adequate cows will be purchased to achieve this objective, the higher the yield per cow, the less that will need to be purchased, and vice versa. In other words, the budget yield for the whole farm comes first, the actual number of cows to be kept comes second.

4. The youngstock management/strategy has also been examined closely: at the end of December 1993 there were 40 in-calf heifers on hand, and 20 bulling/yearling heifers. If the previous management strategy was continued, no more of these would be served until September, ie to calve in June 1995. It is agreed that ten of these will be big enough to serve in February/March, to calve in November/December, and this strategy is to be adopted. Hopefully, the remaining ten will be big enough to serve in April/May to calve in January/February 1995.
5. The service strategy has also been discussed in detail. Most of the cows are of modest genetic potential so a beef bull will continue to be used on the majority of the herd but the Hereford bull is to be sold after he has served the 20 heifers referred to above and replaced by a young Holstein bull of high genetic potential. He will be used on the next crop of replacement heifers, as these are considered to be of higher genetic potential than the cows. It is planned to use him on the heifers for the next 2 years and hopefully, in due course, he will become a stock bull.

   The budget price for the bull is £2,000 and it is planned to purchase one with similar potential to that of 'Moet bulls'. A hundred straws from proven bulls are to be purchased for use on the 60 best cows and hopefully this will result in due course in there being 25 to 30 heifers to enter the herd from the cows, plus 15 to 20 from the heifers, that is 40 to 50 per year.

   A Simmental or Limousin bull will continue to be used on the remainder of the herd and to run as a sweeper bull with the whole herd during the summer.
6. Expenditure on fertilisers and other forage costs is to be kept at a similar level to previous years, but it is planned to increase the stocking rate to approaching 2.5 livestock units per hectare, compared to 2.125 in 1993 (see Table 10.3).

   The change in calving pattern will result in more emphasis being placed on producing milk from grass, rather than from conserved forage. It is tempting to start growing forage maize but this is to be delayed for the time being as a new silage clamp would be required.

   The field on which the forage maize would be grown is one of the best fields on the farm and is eligible for Arable Area

Payments. It is in need of re-seeding, so a spring cereal crop is to be grown in 1994. The work will be done entirely by contract except for ploughing and working down ready for drilling. (Note: On many dairy farms this would not be possible due to the heavy nature of the land and would therefore not be attempted.)

Alternative ways would need to be found to increase the stocking rate, such as making a more substantial increase in cow numbers.

The growing of spring barley on this farm is considered a risk, but it is considered this is worth taking in view of the availability of the Arable Area Payment and the fact that this will mean that a good re-seed can be carried out in the autumn. Ideally, discussions on strategy would have taken place 4 to 5 months earlier, so that a crop of winter barley could have been grown instead of spring barley.

The increase in stocking rate could lead to a shortfall of winter feed in a dry year. It is therefore planned to purchase 100 to 150 tonnes of brewer's grains and to ensile these under the silage, 60% under first cut and 40% under second cut.

Table 10.3 Budget stocking rate

		1994		*1993*
		Livestock units		*Livestock units*
Dairy cows	140	140	130	130
Dairy heifers	80	40	80	40
		180		170
Forage hectares		72		80
Stocking rate livestock units per hectare		2.5		2.125

CASH FLOW ESTIMATES

The budget assumptions made in arriving at the proposed farming programme have now to be looked at from a cash flow point of view and the reader is referred to the data shown in Table 10.2. Before discussing this data, attention is drawn to the way in which this cash flow data is set out, as it has been found in practice this is a very good way of preparing a cash flow budget. The following points are made in relation to its layout and presentation:

1. The actual results for the previous year are shown in column A, and the budgets for the current year are shown in column F. The budget totals shown in column F are then broken down into quarters in columns B, C, D and E.

   Notes regarding the various budget assumptions are set out on the right-hand side of the table.
2. It is emphasised that these are budget targets, and not forecasts of what will happen. The budget objectives are shown in column F and these are then broken down into the quarterly figures, not the other way round. In other words, one starts with the budget objective and looks at the cash flow implications for this objective.
3. The budgets are produced on a quarterly, not monthly basis. Experience has shown that producing monthly data confuses the picture, in other words, one 'cannot see the wood for the trees'. Bankers, unfortunately, tend to like to see the information broken down into monthly estimates. This is OK once the final plan has been arrived at, but preparing monthly data is very time-consuming and laborious, even with a computer, when one is trying to arrive at what the business plan should be, taking into account its cash flow implications rather than producing a detailed forecast, and this is what we are setting out to do in Table 10.2.
4. Attention is also drawn to the fact that the cost of buying-in concentrates is shown at line 2. In other words, line 3 shows the budget MoC.

   In this instance, the budget target is to produce an MoC in 1994 of £140,000 compared to an actual MoC in the previous year of £125,000, an increase of £15,000. (Note: These are cash flow estimates and will therefore be in relation to the milk actually produced in the year ending November, not December, as the income is received one month in arrears.)

   It is interesting to note that the budget MoC for the first two quarters is in the region of £10,300 per month, but for the second two quarters it is in the region £13,000 per month. This reflects the pattern now found on many farms due to the influence of seasonality payments.
5. The data shown in column A indicates that this particular farmer had a good cash flow result in the year ended December 1993, reflecting the improvement in milk price, calf and cull cow prices already referred to earlier in this chapter. During this 12 month

period he reduced his bank overdraft by £20,000 and at the same time reduced his bank loan by £1,400.

6. As mentioned earlier, it is assumed that most fixed cost items in 1994 will be similar to 1993, and these assumptions are detailed in lines 15 to 20. Note that an increase of £9,500 is shown at line 17 to allow for the wages to be paid to the son as well as the increase in other wages due to inflation.

   The proposed purchase of dairy cows and a bull in the quarter January to March leads to a substantial cash deficit in this quarter. The cash flow estimates, however, suggest that these cows can be paid for by the end of the year by which time the overdraft is expected to be £2,600 less than at the start of the year, plus £2,400 loan repayments equals a total reduction in bank borrowing of £5,000.

7. Note the question marks shown at lines 4 and 13 for the possible sale and purchase of dairy cows in the quarter October to December and a similar question mark at line 25 for possible investment of funds in machinery and equipment.

   Decisions about whether or not to go ahead with the ideas previously outlined in this chapter will be made at that time in the light of actual results compared to budget.

   Finally, before leaving the discussion of cash flow estimates, note the contingency shown at line 7, for a possible shortfall in the MoC of £5,000. These are the figures that are to be presented to the bank and in effect are based on an MoC of £135,000. The management objective is to produce an MoC of £145,000. This latter figure is the one against which progress during the year will be measured and is the figure that will be discussed in detail with the herdsperson at monthly meetings when monitoring progress against the budget target.

BUDGET CHANGE IN LIVESTOCK NUMBERS AND LIVESTOCK RECONCILIATION

This is set out in Table 10.4, based on the strategy discussed earlier, to purchase an additional 15 cows in the quarter January to March, and to aim to transfer 40 heifers into the herd before the end of December. Some of these heifers may not calve down before the end of December, so this reconciliation shows 35 heifers being transferred into the dairy herd, not 40, to allow for this possibility.

This livestock reconciliation suggests that the number of heifers on hand at the beginning and end of the year should not change, but the number of cows on hand should increase by 18.

Table 10.4 Livestock numbers reconciliation

		Dairy cows		Dairy heifers
1.	Dairy cows and heifers:	*Dairy cows*		*Dairy heifers*
	No. at start	129		80
	ADD purchases	15		—
	Transfers in (newly calved heifers)*	35	heifer calves*	35
		179		115
	LESS Sales	30		—
	Casualties	2		—
	Transfers out	—		35
		32		35
	Number at end	147		80

* Target is to transfer in a total of 40 but budget is conservative and only allows for 35.

2.	Calves:	
	Calves born alive to cows	90
	Calves born alive to heifers	32
		122
	Heifer calves retained	35
	Casualties	7
	Sold	80
		122

BUDGET CASH FLOW PROFIT BEFORE DEPRECIATION

Having worked out the probable cash flow and the expected change in livestock numbers, we are now in a position to work out what the likely profit will be for 1994 before adjusting for changes in creditors and debtors and depreciation—see Table 10.5.

Table 10.5 Budget cash flow profit before depreciation

				£
A.	*Budget trading cash surplus*			
	Budget trading receipts net cost of concentrate feed for dairy herd (Table 10.2, line 8)			168,000
	LESS Budget trading payments (Table 10.2, line 23)			140,000
			A	28,000
B.	*Budget change in stock valuation*			
	Dairy cows	Increase 18 @ £1,000		18,000
	Youngstock	No change		—
	Bull	Limousin replaced by Holstein		+1,000
			B	19,000
C.	*Budget cash flow profit before depreciation A*			
			A+B	47,000

MILK QUOTA PURCHASE OPPORTUNITY

Opportunism is a theme of this book. This chapter was written in November 1993 and updated in May 1994. Nationally, we have finished the year over quota and the Milk Marketing Board have worked on the assumption that the threshold may be no more than 2%, although in practice it is still hoped that it may be closer to 3%.

The price of leasing milk quota has firmed in recent weeks and is now in the region of 7p per litre, 1p more than that shown in the budget for the example farm.

Milk quota purchase prices, however, have eased so it has been decided to take the opportunity/risk and purchase 100,000 litres of 3.95% butterfat quota for £42,000 (42p per litre). The bank have been approached and have agreed to the above based on a repayment programme over a period of 7 years. It has been agreed that £500 per month will be transferred from the current account to the milk quota loan account with interest charges being debited to the current account.

Time will tell whether this proves to be an opportunity well taken or a risk that should not have been taken.*

* July 1994. So far so good. Purchase prices for 3.95% quota have risen to 47p, and leasing charges have increased to 8p.

CHAPTER 11

Implementing the Farm Plan or 'Making it Happen'

In Chapter 10 we discussed the preparation of our plans and objectives for the year; in other words, we have been deciding what to do. We now have to move on to the second phase, put the plans we have prepared into operation, and achieve the budget objective.

ATTAINING THE BUDGET HERD YIELD

In the previous chapter it was agreed that the management objective is to increase the yield by 100,000 litres compared to that achieved in the previous year. It is hoped to achieve this in part by increasing the yield per cow, but the most certain way to achieve it is to purchase a few additional cows.

A yield prediction needs to be prepared as described in Appendix 3 to determine the number of cows that need to be purchased. In this connection it is suggested that one thinks like a football manager who has 11 players and two substitutes and brings on the substitutes only if required. If a few cows are purchased at the beginning of the year, cows that are not performing can be discarded towards the end of the year. It is most important 'to keep cow numbers up to budget'. This may seem a simple thing to do, but not having enough cows is the main reason for lack of success on many dairy farms. For example, a farm with which the author has been involved took a decision in January/February 1993 not to purchase ten additional cows, as the manager thought they would not be required. As time progressed it became evident that these cows should have been purchased and ten additional cows were eventually purchased in December 1993 to ensure that the problem was not repeated in 1994. At the time of writing this book (March 1994), these additional cows are in the herd, they are performing well and production is approximately 300 litres per day more than it was a year ago. As the year progresses it is hoped that production will continue to be above that achieved a year ago so that a few cows that have been identified as potential culls can be sold in mid-summer/autumn.

KEEPING COW NUMBERS/FEED SUPPLIES IN BALANCE

Having worked out the number of cows required to achieve the budget yield objective, we now have to try to ensure that adequate feed supplies will be available, particularly in mid-summer and during the winter. The former is important due to the emphasis placed on high prices for summer milk. The latter is important as it will allow a late spring to be negotiated without the need to purchase expensive feed supplies to eke out a shortage.

Running short of feed in April is one of the most certain ways of ensuring that the budget targets are not reached. Most costing schemes commence in April and a poor result in this month tends to depress the results for the rest of the year.

Cow numbers/feed supplies is a basic and fundamental profit factor which must always be borne in mind when considering the merits of the alternative schools of thought in relation to grassland conservation, ie whether to go for bulk or quality.

Whatever system you adopt it is important that, come October, you assess both your winter feed supplies and the targets you set earlier in the year against budget and then make any necessary adjustments to your strategy. However, October may be too late: by the end of July you should have a good idea of your likely feed stocks and this is the time to make decisions such as whether to grow stubble turnips or order additional brewer's grains. If you use sugar beet pulp you should have ordered enough last year to feed in September and October this year, ie before this year's supply becomes available.

The switch in emphasis to summer calving has also made it desirable to have forage maize silage available to feed to cows in the summer months and in recent years considerable interest has also been shown in whole crop cereals. The end of July is the time to start to fine-tune decisions that will ensure a good supply of food throughout the late summer and winter period. This is the time to take decisions to buy in extra food, eg due to a drought, not wait until the end of the winter, hoping that you will have enough.

Where cash flow is tight, this may be a difficult decision to make as the food purchased could eventually turn out to be surplus to requirements. The successful dairy farmer, however, always has a reserve of food supplies available, just in the same way as he has a few reserve cows/heifers to introduce into the herd, if the need should arise.

Assessing the quantity of feed supplies relative to the livestock requirements is not as simple as it might appear: Firstly, the length of the winter period, or date of turnout, is not known to within 30 days and this represents at least 15% of the total requirement. Secondly, the weight of silage in a clamp varies by up to 20% depending on whether one believes there is 0.65 or 0.78 tonne per cubic metre.

This is particularly true on self-feed silage systems. A very useful tip in this case is to keep a simple record of the date each bay is completed. This will allow you to calculate the number of cow days per bay and will be valuable information in the year of shortage. There is no excuse however, for 'running out of silage' except in rare circumstances. The number of days to 'turn-out' should be calculated and the home-produced feed should then be rationed accordingly. On a self-feed silage system this may simply mean that the fence has to be moved 10 cm per day instead of 12 cm. This will reduce silage intake and consequently other feeds will need to be used. Whether this takes the form of purchased concentrates, brewers' grains, home-grown cereals, hay or other purchased feed will depend on individual circumstances.

ACHIEVING THE BUDGET MoC

As mentioned in the previous chapter, the most certain way of achieving the budget MoC is to set a target which is higher than the budget. Monitoring the MoC against this budget target is the most certain way of seeing and checking that the cows and the feed are managed effectively.

It is also useful to take part in some form of group costings scheme but beware the dangers of 'cooking' the results so they look good or drawing the wrong conclusions owing to the lack of authenticated data. Try to avoid drawing too many conclusions from one month's figures; ideally we want to be able to see one month's figures in relation to the year's progress.

When things are going well the comparison of results to budget is very pleasant, especially when the results are better than budget and better than those achieved in the previous year. In these instances it is often difficult to pinpoint exactly why progress is so good and the same is true when things are not going well! This, however, is the time when comparing results to budget, as well as to the previous year, is vital, as it leads to decisions being taken, albeit at times difficult decisions, to ensure the budget objectives are achieved.

An example that comes to mind is a farm that in 1993 was not achieving the progress in total MoC that had been expected. On examining the results in more detail it was discovered that the results in the previous year, ie 1992, had been better than expected as the feed quality/weather was better in 1992 than in 1991. A decision was therefore taken to purchase more cows than had originally been planned. This had an adverse effect on the cash flow in the short term, but the outcome by the end of the year was in line with budget.

LACTATION MONITORING

The monthly data tells you how well you have done but if the results are not as good as expected it does not tell you the reasons why and it also tends to be too late for remedial action.

In the 1970s Ken Slater developed his Brinkmanship Recording System based on the standard lactations of cows and Genus has developed the Herdwatch Scheme, based on similar principles. On many farms the comparison of the actual results of groups of cows according to their month of calving is a vital part of successful herd management. The key factor in this management is the production of a graph which gives a visual illustration of actual trends compared to standard lactation curves (see Figures 11.1 and 11.2). In most dairy herds, calving is spread over a wide period and at any point in time there are dry cows, cows in early lactation, cows in mid-lactation and cows in late lactation. The feed requirements of these cows at differing stages in their lactation vary widely as do their average yields. It is pointless, therefore, to study the average yield for the herd as a whole.

The essence of the Brinkmanship Recording System is that the milk yields of individual cows are recorded weekly on a group basis according to the month in which they calve. The average yield for the group is then plotted and related to feed input and to yield expectations. The plotting of milk yield in this way can reveal some startling results. This is illustrated with data taken from two farms.

Figure 11.1 Lactation curve for March-calving group of cows

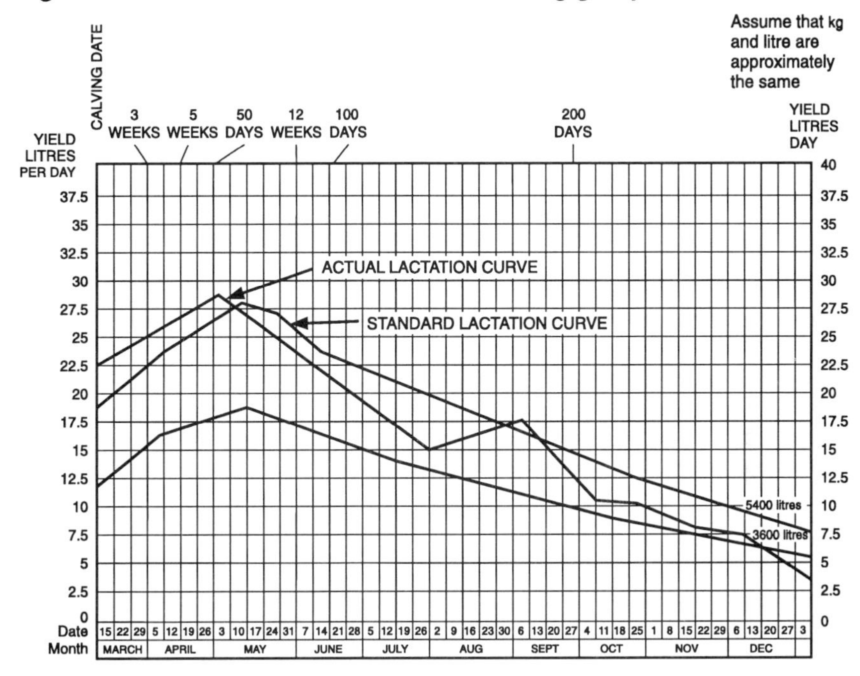

Figure 11.2 Lactation curve for September-calving group of cows

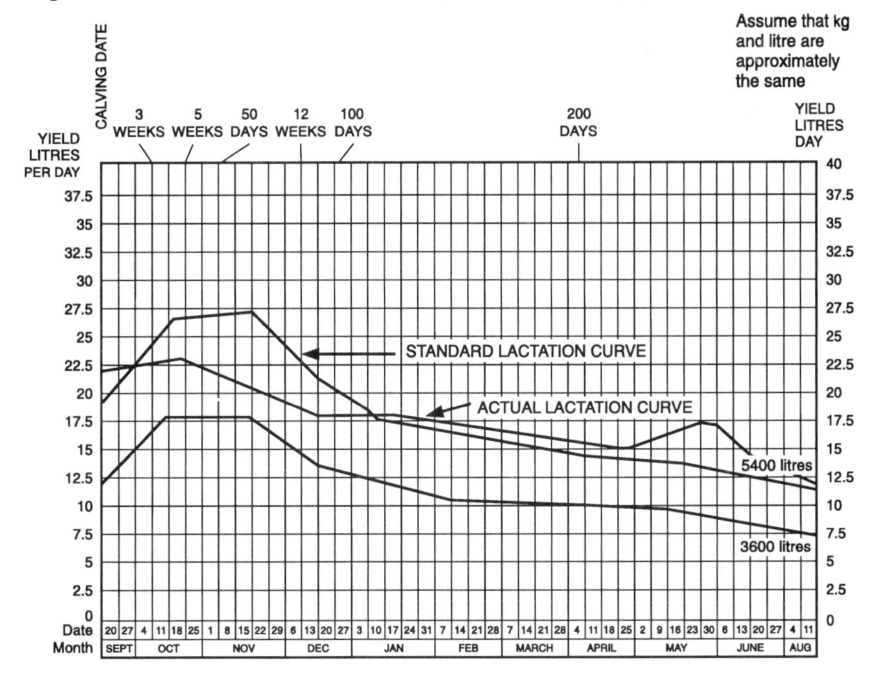

The results have been plotted *in retrospect* and are shown in the lactation charts.

The first is for a March-calving group of cows (Figure 11.1). Milk yield in early lactation showed promise but fell away in June and July due to drought then revived briefly in September with a late flush of grass. In late lactation with the onset of a cold, wet, early winter, production again suffered as the cows were left outdoors too long.

The second is for a group of cows calving in September (Figure 11.2). The peak yield is low as the cows were out at grass in October. There is a small improvement as the cows were brought indoors but the main feature of the lactation chart is the substantial rise on turning out to grass.

These illustrations show what can happen when no records of performance are kept, or to be more accurate, records are kept but not used. The objective of weekly recording is to try to prevent unsatisfactory yields by taking appropriate action before it is too late.

The lactation charts used show 'standard curves' for cows calving in these months. They are a useful guide but are not appropriate for high-yielding herds as the standard lactation is based on national results and these include herds that are 'mismanaged'.

If you have a high-yielding herd you will be looking for a flatter curve than those shown and for a much higher peak yield, probably in the region of 40 to 50 litres, with a target for the whole lactation of 8,000 to 9,000 litres.

(Note: The Brinkmanship System has become associated in peoples' minds with 'feeding to yield' with undue emphasis on feeding concentrates. The Brinkmanship name, however, was coined with exactly the opposite objective in mind, the idea being to monitor the lactation curve with a view to reducing the concentrate input to as low a level as possible without a loss in yield. It is a vital tool in this aspect of dairy feed control, as during the summer period the key to success is the ability to manage forage, ie mainly grass and other conserved foods, to ensure that the yield is at the level it should be without feeding concentrates).

Lactation monitoring is also vital where complete diet feeding systems are practised as a means of ensuring that milk production is on line with what it should be.

Studying the lactation curves on your farm is an excellent way of discovering how your cows perform according to the month of calving. Experience over the years has shown that it is very difficult to manage cows that calve during the summer/autumn before being housed and achieve a satisfactory yield. This statement could not be made if the author had not been closely involved with managing various groups of cows and discovered this at times to his cost, that is the money in his own pocket. Experience has also shown that it is difficult to overcome the problem illustrated in Figures 11.1 and 11.2. This book is essentially about business management, but in view of its importance a few words are now said about the management of the cow in relation to its standard lactation curve.

The lactation curve can be conveniently divided into two parts:

1. Early lactation: Calving to peak yield

This is a vital time as during this period the pattern is set for the whole of the lactation. It is generally accepted that the lactation yield of a cow is equal to its peak yield × 200 and that the lactation of a heifer is peak yield × 240. It is therefore vital that a group of cows calving in, say, September should average 35 litres per day in October to achieve a lactation yield of 7,000 litres. If this peak lactation yield is not achieved action has to be taken to find out the reason why and appropriate remedial action taken.

(Note: As mentioned above, in practice it is often very difficult to attain satisfactory peak yield from cows calving at grass in August/September but it is much more simple to achieve with cows calving down in November/December, ie after housing. This has led to a practice on many farms to house autumn calving cows before they calve down and this action may need to be taken on your farm.)

At, or just after, attaining peak yield most cows are served with a view to getting them in calf for the next lactation. This is a very critical time in the management of the dairy cow and leads to the need to keep adequate breeding and veterinary records, to discover whether or not there is a problem getting cows in calf.

During this period, ie from 6 weeks to approximately 12 weeks after calving, considerable attention must be given to ensuring the cow is properly fed so that she conceives and calves down to commence her next lactation, hopefully within a period of 365 days. In practice however, it has to be accepted that with some exceptionally high yielding cows, 365 days lactation is not attainable.

2. Mid/late lactation

During this period the emphasis changes from trying to ensure that the cow is fed adequate high-energy feeds to attain a high peak yield, to one of substituting cheaper bulk food/forage crops and grazed grass for more expensive concentrates. During this period lactation monitoring takes on the significance for which the term 'Brinkmanship' was originally coined, ie to cut the level of concentrates with a view to increased reliance on grass and forage. It is at this stage in the cow's lactation that the cow becomes a much better measure of the value of forage than any chemical analysis, or *in vitro* (test tube) method of valuing such forages as silage.

To make this assessment in, say, January, a reduction would be made in the level of concentrate feeding to the cows that calved in August/September to see if there is any drop in yield. Over a period of several weeks one arrives at a very good indication of the contribution that the silage/bulk foods can make to milk yield and can then feed the rest of the cows, ie the later calving groups, accordingly.

CONTROLLING FEED COSTS

Accurate records must be kept of the quantity of concentrates fed on the one hand to the dairy cows, and on the other to youngstock and beef cattle. In this connection, it is suggested that management should set out to make sure that what is fed is the same as what 'it is intended to feed'. This can only be achieved by keeping detailed records.

On farms where all the concentrates are purchased this is relatively easy but it becomes much more difficult if home-grown cereals are used and/or if straights are purchased to be used in a complete diet system.

Surveys have shown that parlour feeders are not always accurate in dispensing feed and these records will quickly pin-point this problem so that appropriate action can be taken. The availability of feeder wagons with weigh cells allows a check to be made on quantities of bulk feed as well as concentrates.

A feed recording system needs to be set up along the lines illustrated in Table 11.1. This illustration shows that the amount of high-energy concentrates intended to use was 6,500 kg, but the actual amount used is 7,000 kg.

The reason for the discrepancy of 500 kg needs to be investigated, as does the discrepancy in the use of sugarbeet nuts, which is 50 kg more than planned. This aspect of the management of the dairy herd could be very much in the hands of the herdsperson, as well as the farmer, and both need to be involved in the discussion to

Table 11.1 Feed recording example

	Concentrate type	
	High energy kg	*Sugar beet nuts kg*
Opening stock	5,000	2,000
Deliveries	5,000	2,000
Total available	10,000	4,000
LESS closing stock	3,000	3,500
Total used	7,000	500
Amoung expected to use	6,500	450
Discrepancy	500	50

determine the reasons for the discrepancy and to decide what can be done about it.

(Note: On farms where home-grown cereals are grown, an estimate should be made immediately at harvest time of the tonnage harvested and this should be compared, as the winter progresses, to the amount actually fed according to the feed records. The author is very well aware of many examples where detailed feed records have been kept that do not tally with the tonnage harvested. A very frequent comment on the results for mixed dairy/arable farms is: 'The cereal yields would appear to be quite modest according to the amount shown as being fed to livestock, but in practice the amount shown as being recorded to livestock could have been underestimated'.)

CONTROLLING THE CALVING INTERVAL

Mention has already been made of the importance of getting cows in calf and the attainment of a good calving index. The importance of the latter, however, does tend to be given too much importance by specialists whose main concern is with this aspect of management.

Claims are often made that a loss of, say, seven days in the calving interval is equal to £X loss of profit per cow per year. This claim, however, is often made on the erroneous assumption that the cow concerned would be dry at 305 days. This is no longer true on many farms as the cow is capable of milking for a further 30 to 60 days with a very satisfactory margin in the extra 30 to 60 days of her lactation.

It is, however, still very important to ensure that most cows calve down as and when planned, and very few farmers will not aim to calve cows down at 365 days. Loss of profits, however, occurs if a rule-of-thumb decision is taken that a cow that does not calve within, say, 390 days should be sold as a barren simply to achieve a good calving index. This decision is not justified, particularly when the price of replacements is high. The correct management decision should be to plan to get her in calf, and if necessary, sell her as a down-calved cow in the following year.

It is also important to mention that in a quota situation the effect of a poor calving interval could well be that less cows have to be kept, on average, to produce the planned amount of milk, as the extra production from a given number of cows would be penalised.

The calving index becomes vital where farms aim to calve all their cows over a very short period, eg in the spring, and at the same time have a policy of feeding virtually nothing but grass to their cows. This strategy in itself makes it difficult to get cows in calf as they tend to be underfed at conception and problems occur with such factors as mineral imbalances.

This is particularly noticeable in parts of Southern Ireland where it is normal practice to run *two* beef bulls with a herd of, say, 80 cows to try to ensure they get in calf. These cows are dry by 305 days so a substantial loss of income does occur if they are not in calf.

ENSURING CALF OUTPUT EXCEEDS HERD DEPRECIATION

Most dairy farmers take part in some sort of costings which determines their MoC per cow and the margin from the total herd. Very few, however, take part in costing schemes which monitor the trend in the calf output, relative to the herd depreciation in the way that they should.

This aspect of dairy business management is particularly important at the present time as the difference between a newly-calved heifer and a cull cow is in the region of £800, ie £1,250 less £450. Five calves at £160 per head, or four at £200 per head, are required to offset this shortfall in the price of selling a cull cow compared to purchasing its replacement. In this connection, it is important to note that a heifer purchased as a down-calver will not produce a calf until it has been on the farm for a year.

If 25 cows out of a herd of 100 are sold as barren, this leaves only

75 cows to calve. After allowing for a 5% mortality and a calving interval of say 390 days, the number of calves born alive is only in the region of 65 to 70. In other words, there are less than three calves to sell for each heifer entering the herd and there is virtually no chance that the value of these three calves will equal the difference between the sale value of the cull cow and its replacement.

It is therefore normally good practice to try to ensure that heifers are bought in as in-calf heifers. The value of the calf in most instances will outweigh the risk of calving losses and the down-calved heifer is also likely to perform more effectively if she calves down on the farm.

Another way to overcome this problem, as mentioned when discussing the calving index, is to ensure that as many cows as possible are got in calf. A down-calved cow and her calf will then be available to offset the cost of a purchased replacement, if this is thought necessary, eg to continue with a tight calving pattern.

CONTROLLING CASH FLOW

Experience has shown that information set out as shown in Table 11.2 is an excellent aid to the management of cash flow. The actual cash flow results and the budget are set out on a cumulative basis.

Set out in this way it is possible, for example, to look at the cash flow after, say, 6 months, compare the results at that time to the budget, and at the same time look how progress is being made to achieve budget target by the end of the year.

When studying this data it is important to note whether or not any variance is simply due to timing of payment or sale. One often finds, for example, that the payment for feed (line 2) may be below budget but this is simply because a cheque has not been written out for the feed.

From a cash control point of view, therefore, it is recommended that the cheque should be written out before the end of the month and entered on the cash flow schedule, shown in Table 11.2, put back in the drawer and then posted just before the milk cheque is due in the following month. This cash flow sets out to measure what the overdraft would be if all the cheques that should be paid were paid. Consequently, the overdraft figure shown at line 50 is the 'book overdraft', which is the actual overdraft plus cheques drawn, not presented.

The key to effective management is careful control of all inputs and the ideal situation is to find that after ten months the actual

Table 11.2 Example of farm cash flow compared to budget for year ending 30 April 1994

	DETAILS		A	B	C	D	E	F	G	H
			ACTUAL			BUDGET 3 MONTHS	ACTUAL			BUDGET 6 MONTHS
			1 M	2 J	3 J		4 A	5 S	6 O	
TRADING ITEMS	Milk sales	1	18304	35162	51812		76540	105016	126007	
	Less concentrate cost	2	3285	5179	8548		11696	19559	25969	
		3								
	MoC	4	15019	29983	43264	40000	64844	83457	100038	95000
	Cull cows/heifers	5	(4) 2327	2327	2327	2000	(5) 2618	2618	2618	4000
	Surplus cows/heifers	6								
	Calves	7	(7) 1213	(11) 2480	(34) 6522	3000	(56) 10698	(63) 11927	(69) 12866	7500
	Beef stores	8	(12) 3787	3787	3787	4200	3787	3787	3787	7700
	Bull beef	9								
	Other cattle	10			956	–	956	956	956	800
	Contract rearing	11								
		12								
		13								
	TOTAL TRADING Receipts	14	22346	38577	56856	49200	82903	102745	120265	115000
Capital Items	M/c & Equipment	15								
	Grants	16						9040	9040	
	New Loans	17								
	Milk quota compensation	18	2099	2099	2099	1200	2099	2099	2099	1200
	Tax rebate	19							709	
	V.A.T.	20	3391	3391	3391		9525	9525	9525	
	(A) TOTAL RECEIPTS	21	27836	44067	62346	50400	94527	123409	141638	116200
TRADING ITEMS	Livestock purchases	22						2000	2000	2400
	Youngstock feed	23	242	919	1406	1000	1406	1406	1958	2000
	Vet/Med/AI	24	1166	1802	2382	2500	2826	3431	4464	5000
	Livestock sundries	25	351	684	945	1000	1480	1798	2185	2500
	Fertiliser	26	4319	4319	4319	6000	4319	4319	7498	8000
	Milk quota lease	27			3008	2000	3008	3008	3008	4500
	Seed/Spray/Sundry	28	3	215	346	500	1672	1688	1773	3500
	Bulk food	29		5859	6462	5000	9105	10098	8068	6000
	Cattle keep/grazing	30	2040	2040	2040	2100	2040	2040	2040	2100
	Regular Labour	31	4300	6450	6450	8700	10750	12900	15050	15400
	Contract	32	1200	4334	7094	9000	7847	9537	10639	15000
	M/c Repairs	33	886	1119	1279	1000	2314	3446	4393	3000
	Fuel & Power	34	576	1136	1136	2000	2154	3289	3622	4000
	Rent & Water	35	1360	2879	4900	4700	6422	7870	9841	9400
	Property Repairs	36	264	1716	1670	1000	1670	1670	1702	2000
	Misc. Inc.Insurance	37	500	898	951	1000	1021	1511	1847	2400
	Interest Charges	38		2587	2587	3000	2587	5482	5482	6000
	TOTAL TRADING Payments	39	17406	36957	46975	50500	60708	75525	85570	93200
Capital and Private Items	Private Drawings & Tax	40	700	1581	3231	4200	4231	5231	16231	7200
	M/c & Equipment /HP	41	297	594	891	1200	1188	1188	1250	2400
	Land and Improvements	42	7014	7014	7014	7000	7014	7014	7014	7000
	Loan Repayments–Bank	43	300	15600	15900	15900	33300	33600	33900	31800
	Sprecial drawings	44						10287	10287	10000
		45								
	V.A.T.	46	3888	5473	6135		7788	10833	12914	
	(B) TOTAL PAYMENTS	47	29605	67219	80416	78800	114229	143678	169614	151600
	(C) SURPLUS/DEFICIT	48	(1769)	(23152)	(17800)	(28400)	(19702)	(20269)	(27976)	(35400)
	Op. Bank Bal.(Overdraft)	49	(32892)			(32900)				(32900)
	Cl. Bank Bal. (Overdraft)	50	(39968)	(56044)	(50692)	(61300)	(52594)	(53161)	(60868)	(68300)

outlay on cost items is no more than the budget for nine months. If at the same time the income after nine months is in line with the budget for ten months the business, to say the least, is on the road to success.

It is often said that it is not possible to predict in dairy farming what is going to happen but it is surprising how often, when a

	I	J	K	L	M	N	O	P	
	ACTUAL			BUDGET 9 MONTHS	Actual	Budget March and April	Revised 12 month Budget	BUDGET 12 MONTHS	Variance
	7 N	8 D	9 J		10 F				
1	147701	164057	179544		202171	37000	239111	231000	+8111
2	26471	29972	34312		36153	10000	46153	51000	(4847)
3				135000				180000	
4	121230	134085	145232		166018	27000	193018		+13018
5	2618	(18) 8046	(19) 8331	8000	8331	7000	15331	13000	+2331
6					(1/4)				
7	(80) 14551	(90) 15676	(103) 18011	12000	19586	6000	25586	18000	+7586
8	3787	(12) 3922	3922	7700	3922	–	3922	7700	(3778)
9									
10	956	956	956	800	956	–	956	800	+156
11			1200		1200	600	1800		+1800
12									
13									
14	143142	162685	177652	163500	200013	40600	240613	219500	+21113
15									
16	9040	9040	9040	5000	9040	–	9040	5000	+4040
17									
18	2099	2099	2099	1200	2099	–	2099	1200	+899
19	709	709	709		709	–	709		+709
20	16305	16305	16515	16515	20754	–	20754	20754	
21	171295	190838	206015	186215	232615	40600	273215	246454	+26761
22	2000	2000	2000	2400	2000	20000	22000	22400	(400)
23	3180	3180	3180	4500	3889	1000	4889	6000	(1111)
24	5417	6735	7822	7500	8845	2000	10845	10000	+845
25	2709	3032	3406	4000	3818	1000	4818	6000	(1182)
26	7498	7498	7498	8000	7498	5000	12498	12000	+498
27	6128	6128	6128	4500	6128	–	6128	4500	+1628
28	1773	1773	1773	3500	1773	–	1773	4000	(2227)
29	11825	11825	11825	6000	11825	–	11825	6000	+5825
30	2040	2040	2040	2100	2040	–	2040	2100	(60)
31	15050	17308	21608	22100	21608	6500	28108	28800	(692)
32	15152	15152	15322	16000	15495	1000	16495	16000	+495
33	5028	5803	6515	5000	6916	1000	7916	7000	+916
34	4231	5095	5746	6000	6333	1500	7833	9000	(1167)
35	11201	13463	15143	14100	16595	3500	20095	18900	+1195
36	1768	1885	2172	3000	2749	1000	3749	4000	(251)
37	2017	2822	2897	4400	2885	3000	5885	6400	(515)
38	5482	7746	7746	9000	7746	2000	9746	12000	(2254)
39	102499	113485	122821	122100	128413	48500	176643	175100	+1543
40	16731	18231	22011	11400	22511	3000	25511	14400	+11111
41	1339	1339	1339	4200	1339	–	1339	6000	(4661)
42	7019	8019	8019	7000	8019	–	8019	7000	+1019
43	34200	34500	34800	32700	35100	600	35700	33600	+2100
44	10287	10287	10287	10000	10287	–	10287	10000	+287
45	2448	2448	2448		2448	–	2448		+2448
46	14767	16032	17363	16515	17903	3000	20902	20754	+148
47	189290	204341	219088	203915	225750	55100	280850	266854	+13996
48	(17995)	(13503)	(13073)	(17700)	+6865	(14500)	(7635)	(20400)	+12765
49				(32900)		(26027)	(32892)	(32900)	
50	(50887)	(45395)	(45965)	(50600)	(26027)	(40537)	(40537)	(53300)	

determined effort is made to control costs, to find these can in fact be kept in line with budget.

The extent to which this is true can be judged by looking at the data in Table 11.2. This information has not been made up; it is in fact the actual results on a farm with which the author is very much involved.

(Note: The actual results as previously stated are set out on a cumulative basis. They are also tabulated on a separate sheet on a month by month basis so that if one wishes, one can look at the results in a particular month in more detail.)

Let us look now at the actual results achieved by this farm during the year ending 30 April 1994.

The actual trading payments (line 39) after 9 months, at £122,821 are virtually identical to the budget target of £122,100.

The expenditure on bulk food at £11,825 is £5,825 more than budget but this is cancelled out by saving on several other items, eg youngstock feed and seeds, sprays and sundry crop costs, and the outlay on purchased concentrate feeds is also below budget.

Total trading receipts net concentrate feed (line 14) at £177,652 are £14,152 more than budget. The MoC (line 3) is £10,232 more than budget, and receipts for calves (line 7) are £6,011 more than budget.

Receipts for beef stores at £3,922 are £3,778 below budget, but this is simply due to a decision taken not to sell ten of the store cattle but to carry them through to be sold in the following year.

Substantial loan repayments have been made to the bank as planned, actual £34,800, budget £32,700.

Advantage has been taken of the favourable trend in the cash flow to increase private drawings. These, including tax payments, are £10,611 more than budget but despite this, at the 9 months stage there is an actual cash deficit of only £13,073 compared to a budget of £17,700.

The business is performing well reflecting a combination of an upturn in milk and calf prices, coupled with achieving actual physical performance targets in line with budget.

Month 10 (February) was an excellent month from a cash flow point of view. There was a cash surplus of £19,938, giving an actual overdraft at the end of February of only £26,027 compared to £45,965 at the end of January.

Having made it happen and achieved a good cash flow result, now, ie 2 months before the end of the financial year, is the time to prepare a budget of the likely result for the whole year and to take steps where appropriate that will minimise the potential tax liability.

These budget estimates are set out in columns N and O on Table 11.2.

Allowance was made when the budgets were prepared for the possible purchase of replacement dairy cows in this period and a decision was taken to go ahead with this as originally planned.

At the same time, a draft budget was prepared for the year ending April 1995, but that is another story.

MONITORING LIVESTOCK NUMBERS

A good cash flow could be at the expense of a reduction in livestock numbers, or alternatively at the expense of not increasing the numbers as planned. A regular note and reconciliation of livestock numbers needs to be kept, as illustrated in Table 10.4 page 139.

This allows a comparison of the numbers on hand at the present time to be made, either to the number on hand at the start of the year, or to the number budgeted to be on hand at the end of the year. Very often a variance in the cash flow is due to there being either more or less cattle on hand than was budgeted.

COMPARISON OF ACTUAL MANAGEMENT ACCOUNTS TO BUDGET

The final test of management is whether or not the budget profit objectives have been achieved. The objective should be to complete the accounts within three months of the accounting year, ie by the end of March if the trading year ends in December, and at the same time have the budgets available to discuss for the coming year. It is often necessary to prepare draft budgets/strategy for the coming year before these accounts are available.

It is surprising how often a hiccup can occur in these vital three months. Draft budgets, therefore, should be prepared within one month of the start of the budget year and modified later if necessary in the light of the actual results for the previous year.

Bank managers like to be kept informed of the progress of a business and it is suggested that the annual review of this progress should be timed to coincide with the production of these management accounts and budgets, ie approximately three months after the end of the accounting year.

To return to the example farm discussed in Table 11.2, it is May 1994, there has not yet been time to prepare the accounts for 1993/94 but a draft budget has already been prepared and submitted to the bank manager.

The budget objective is to produce a similar milk yield and MoC to that achieved in the previous year. Substantial loan commitments were made in 1993/94. Very little capital expenditure was incurred in either 1993/94 or 1992/93. Substantial investments are proposed in capital items for 1993/94 and it is hoped to fund these out of cash flow.

CHAPTER 12
Milk Quotas

HISTORICAL/POLITICAL BACKGROUND

Milk quotas were introduced to the United Kingdom on 1 April 1984, based on the milk production achieved in 1983. 1983 was a particularly difficult year for most dairy farmers as it was a very wet season, with production on many farms being lower than normal. This led to great controversy when milk quotas were first introduced as many farmers felt their allocation of quota to be unfair and this bitterness is still felt to the present day.

There have been many changes to the quota regulations since they were first introduced. Appendix 4 details the main features of the quota system as it applies to the United Kingdom. The text and tables shown in this Appendix have been taken from the Milk Marketing Board's publication *Dairy Facts and Figures*, 1993 edition. The reader who wishes to familiarise himself/herself with the various regulations should now read Appendix 4 and note in particular how difficult it is to arrive at the amount by which the United Kingdom is over quota and the problems that are involved in deciding which purchaser groups are responsible for paying the levy.

Attention is also drawn to the significance of the butterfat base introduced in 1987/88, and the new quota regulations brought in as from 31 March 1993. In this section on the new regulations there is a note to the effect that the UK regulations from 1994/95 are likely to allocate unused quota to over-quota purchasers in proportion to the purchaser's total quota, *not* in direct proportion to their contribution to the UK excess. This in fact is the case and has important implications in the post-MMB era.

No cuts in milk quota have been imposed since 1 April 1992 and this lack of cuts has coincided with a period of prosperity for dairy farmers.

Looking ahead to the future, it has been agreed by the EC that milk quotas will stay until the year 2000 but their future beyond that date is uncertain.

A cut of 1 to 2% was expected during the current year, ie 1994/95, but at the time of writing, this has not been imposed, possibly reflecting the influence of GATT. Political decisions are perhaps being taken to allow milk production to increase, so that this leads to a fall in prices, bringing the price of milk within the EU close to world market prices, when the GATT proposals are introduced. Time will tell.

Whether or not milk quotas will exist in their present form after the year 2000 is debateable.

Mrs Gillian Shephard, the Minister for Agriculture at the time, made the following statements at a recent RABDF (Royal Association of British Dairy Farmers) conference.

> 'The Government's ultimate objective is to get rid of production controls in the dairy sector and to expose the dairy industry throughout the European Community to free market forces.
>
> Unfortunately, we are not going to achieve this overnight. So we must make the best of the present arrangements and minimise the burden that the quota system places on the industry.
>
> First, we have to avoid any further arbitrary cuts in our national quota. Second, we must ensure that our national quota is fully utilised and that the operation of the quota system doesn't impair the development of fair competition under the new milk marketing arrangements.
>
> Finally, it seems to me that it would make a lot of sense to allow quota to migrate to those parts of the Community, like the UK, where milk can be produced more efficiently.'

This is the Government's view; we as producers have to make decisions in terms of what we expect might be the outcome from the year 2000 onwards, particularly when contemplating the purchase of milk quota, as discussed later in this section.

The existing arrangements for suckler cow and sheep subsidy may be a guide as to what may happen. In the case of sheep and beef production there is a quota on subsidies, not a quota on production, and a similar system for milk could perhaps be introduced after the year 2000.

ADMINISTRATION

The administration of the milk quota system was undertaken by the Milk Marketing Boards of England and Wales, Scotland and

Northern Ireland from 1 April 1984 until 31 March 1994, but as from 1 April 1994 it is administered by the Intervention Board.

The Intervention Board is now responsible for providing data to producers and purchasers groups on the actual production compared to quota and is also responsible for policing the quota leasing and other transfer forms relative to milk quota.

This change in the administration of milk quota coincides with changes in various regulations relating to milk quotas (see Appendix 4). A most important principle that has been accepted is that of 'inter-purchaser leasing'. This means that a producer supplying milk to, say, Milk Marque can sell or transfer his quota to a producer who supplies a different purchaser such as Northern Foods.

It is also proposed that the last date for leasing will be fixed at 15 December and not changed each year as at present, but changes have been made to the rules before and a change in this rule is not beyond the bounds of possibility.

The new proposals for the transfer of quota allow the sale of quota without land. No changes have been made in the 'threshold mechanism' which is described in more detail later.

Finally, and perhaps most important, it is now proposed that a vendor will be able to sell milk quota 'clean' (ie unused quota) at any time during the year.

Under the arrangements prior to 1 April 1994 milk quota could only be sold based on its percentage used on its transfer date. For example, at the end of January a farmer could sell his quota 80% used to a purchaser transferring, in effect, 20,000 litres of clean (unused) quota and 80,000 that have been used, to the purchaser.

This change in the regulations has important implications for the trading activity that can be expected at the end of the 1994/95 and subsequent quota years.

During recent years it has become normal practice for the MMB to produce a quota profile showing actual production against quota, as shown in Figure 12.1. This quota profile has been produced in recent years in September based on the actual production to the end of August and the MMB's best estimate of the production profile for the remainder of the year to finish the year exactly on quota.

Similar profiles have been produced by the other milk marketing boards and at the end of the year the amount of superlevy payable has depended upon the extent to which the UK as a whole was above or below quota. (See Appendix 4, item 3, 'Levies on wholesale sales'.)

The production of the profile for the UK as a whole is now the responsibility of the Intervention Board and it is likely to experience considerable difficulty in collating the data from what is expected to

Figure 12.1 Estimated weekly wholesale output and quota (adjusted for butterfat)

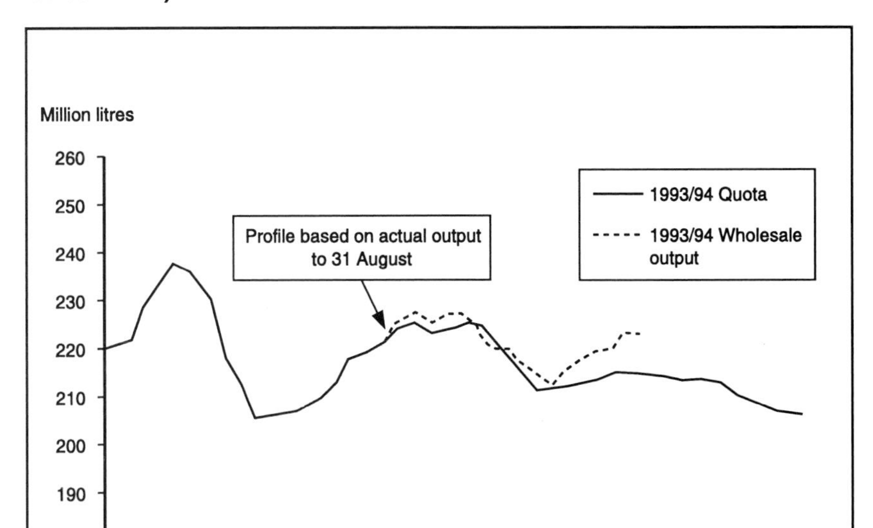

be a substantial increase in the number of purchasers. Individual purchasers will also need to produce similar information for their own individual members and one measure of their success will be how competent they are at putting this into practice.

SUPERLEVY AND THRESHOLD MECHANISM

The existing threshold mechanism has been in operation since the quota year 1988/89 and as mentioned earlier is continuing under the auspices of the Intervention Board and will be calculated in relation to the production in the UK as a whole.

Under the current rules, only those producers 'most over quota' (in percentage terms) pay levy. The point beyond which producers start to pay levy is termed the 'threshold percentage'. If the threshold is say 3% all producers who have not produced more than 3% of their quota (butterfat adjusted) will not pay a superlevy. Individual producers whose production exceeds 3% (butterfat adjusted) will pay a levy on the production over 3%. A farmer who finishes up 5% over quota will pay a levy on 2%, ie the amount by which he has exceeded the threshold.

It has been very difficult for the individual producer to assess what the threshold would be with only five main purchasers of milk, ie the five Milk Marketing Boards (England and Wales, Scottish, Aberdeen and District, North of Scotland and Northern Ireland). It will be much more difficult after vesting day when there will be many more purchasers. Before going on to discuss this in more detail we need to look at the decision each individual farmer has to make. He or she has to decide what the threshold level might be, 2%, 3%, 4% or 5%, and arrange their supplies of milk to their purchaser accordingly, knowing that if they exceed the threshold they will pay a fine, at 1993/94 prices of 28.42p per litre, against a potential receipt for the milk of say 20p per litre, ie an effective payment to the purchaser of 8.42p, to pick up the milk and dispose of it.

It is almost impossible to predict what the threshold will be so each individual producer has to gamble or play it safe and supply to his purchasers no more milk than quota and lose out on a potential bonus if there is a substantial threshold. This bonus can be very substantial indeed, as if there is no levy payable the threshold is infinity.

There was no superlevy liability on wholesale producers in either the 1991/92 or the 1992/93 quota year so all producers over quota in these years escaped paying a superlevy.

Table A4.5 in Appendix 4 shows the distribution of producers over and under wholesale quota in 1991/92 and 1992/93. A small number produced more than 20% over their quota and got away with it, ie 249 in 1991/92 and 135 in 1992/93. Nearly 5,000 producers in both 1991/92 and 1992/93 were between 5% and 9.9% over quota, and approximately 1,000 got away with producing between 10% and 19.9% over quota.

In 1993/94, however, it is a different story. The figures are not yet known but it is likely that a number of producers similar to the above will have had to pay very substantial levies.

On balance, a reasonable philosophy in the past has been to aim to be, say, 4 to 5% over quota and be prepared to pay a fine of 1 to 2% if the threshold proves to be low. Looking ahead to the future, however, this philosophy would seem to need to change in view of the fact that it will be possible to sell quota clean at the end of the year. This is likely to result in there being less producers under quota at the end of the year.

The position is also going to be exceedingly difficult at the end of the 1994/95 quota year as the vesting day for the Milk Marketing Board is 1 November. The Milk Marketing Board will retain quota to cover all the milk that it purchases before 31 October, transferring only the net balance to the new purchaser.

This means that the threshold will have to be worked out for each individual in 1994–95 in relation to only 5 months supply. This is best illustrated by taking a producer who owns, say, one million litres of milk quota and produces 40,000 litres more than quota for the whole of the year. This is 4% of one million litres but could represent 10% of the quota available after vesting day of, say, 400,000 litres.

It would seem prudent, therefore, to base one's production in 1994/95 on a very small threshold of, say, 1 to 2%, especially if one is selling milk to a purchaser other than Milk Marque.

MANAGING THE THRESHOLD/KNOCK-ON EFFECTS

Avoiding 'knock-on' effects is a most important principle when dealing with the problems of being over quota. Fine-tuning needs to start early. In an ideal world adequate quota would have been leased before 15 December in the previous year to ensure there is no need for cutbacks, but as mentioned earlier, it can sometimes be difficult in December to decide what the threshold will be, and sometimes the cows milk better than expected increasing the expected over-quota production from say 2 to 5%, particularly if the butterfat base improves.

This happened to a farmer with which the author is closely involved in the 1993/94 quota year. Production finished the year in volume terms exactly in line with prediction, ie between 2 and 3% over quota, but after making adjustments for the much better butterfat base than had been expected the total adjusted quota was nearly 7% more than quota and the farmer faced a substantial superlevy bill. Fortunately, this was overcome by buying in partly used quota.

Let us look at this in more detail, based on a farm with a quota of one million litres: the farmer has taken a decision to produce 4% more than quota, ie 40,000 litres, this decision having been taken at the beginning of December and based on a normal trend in milk yields.

It is then found in practice that by mid-January the cows are milking well with a potential to produce 50,000 litres more than quota and the view now is that the threshold will be only 2%. There is a need to cut back production in the remaining 75 days of the quota year by 30,000 litres or an average of 400 per day.

The ideal way of dealing with this problem is to sell, say, 14 cows/heifers giving 400 litres per day and replace these with 14 cows/heifers due to calve in late March/April. In other words, having a few surplus animals on hand available to sell is part of the strategic plan of the business and goes back to the planning stage described earlier. The cows potentially surplus to requirements are kept if there is an opportunity to produce over quota but are sold if not. The decision has to be made in good time. Delaying until mid-February increases the reduction in yield required per day to 667 litres (30,000 ÷ 45 days). A decision made at the end of December reduces it to 330 litres per day.

Once again this underlines the opportunist nature of making profits with dairy cows and quotas, and the need to rear more not less replacements to take advantage of the opportunity to produce 'over quota' as and when it occurs.

Most businesses, however, have not made allowance for such factors when preparing their plans and have to make decisions regarding one or more of the following:

1. Dry cows off early: This has the advantage that it does not have any knock-on effects but does mean that the most profitable milk* in a cow's lactation, ie that produced in the last 30 to 60 days, is not produced.
2. Reduce yields by cutting the feeding of concentrates: The danger with this strategy is that it has a knock-on effect after the end of March.
3. Sell cows: This is OK in the short term as it produces ready cash, but has a serious knock-on effect if there are no replacements available to calve in late March/April.
4. Accept one has made an error of judgement, carry on producing and pay the levy. This is probably the best thing to do if it cannot be overcome by simply drying cows off early or buying in more quota in good time.
5. Buy part-used or clean quota: This is an excellent way of dealing with the problem, providing that the need to buy is seen before most other producers, as once the majority of producers are aware of the problem the premium value of 'clean' over its base price tends to become uneconomic.

To illustrate the above, let us take the situation in February/March 1994: At this time the base price for milk quota was in the region of 12p per 1% BF (butterfat) or 47p per litre for quota at 3.9% (48p for quota at 4.0% BF (butterfat). This is the price of milk quota that has been fully used and is in effect being purchased for use in 1994/95.

* Assuming ample bulk food supplies are available for these 30 to 60 days.

The price of clean quota, in other words, quota that was 0% used at this time, was in the region of 67p per litre, a premium over the base price of 21p per litre compared to a potential levy liability of 28p.

The levy liability is subject to tax, in other words, can be set off against tax and its net cost after tax at 25% is 21p per litre and at a 40% tax rate is only 16.8p per litre.

It therefore makes economic sense to pay the fine rather than to purchase quota at these inflated prices. Despite this, many producers purchased quota paying the above premium.

(Note: These very high prices were due in part to the decisions taken by most non-milk producers in December to lease out the whole of their quota as at that time they were not expecting that nationally we would be over quota at the end of the year. The fact that they leased out their quota was in effect part of the reason why production finished up over quota as the lessees had the quota available and produced the milk. In previous years, the opposite had been the case; non-milk producers decided not to lease out all their quota, retaining it with a view to selling it clean at the end of the year. Producers did not have this quota available as their hedge, and made cutbacks in production leading to national production in both 1992/93 and 1991/92 being below quota.)

Returning to the question of purchasing milk quota to get over the problem of being over quota. The best way to do this in 1994 was to buy part-used quota. Various lots became available and deals were eventually struck which divided the levy liability on a 50–50 basis between the seller and the purchaser.

In other words, the purchaser paid a premium over the base price in the region of 14p per litre of clean quota, eg 55p, 48p base price plus 7p for 50% used quotas at 4.0% BF.

BUYING AND LEASING MILK QUOTA

The object now is to discuss the buying and leasing of milk quota in relation to what is a theme of this book, namely that a dairy farmer needs to make the quota fit the farm, not the farm fit the quota.

From 1985 to 1989 it did not make much difference to the profit of the farm whether milk quota was leased or purchased as the annual cost of leasing quota was very similar to the interest on the capital invested. Prices fluctuated during that period but tended to remain within the region of 30 to 35p per litre (excluding the inflation in price at the end of the year for clean quota) and the cost of leasing in milk quota was in the region of 5p per litre.

Interest rates at this time were high, in the region of 15%, giving an interest charge on milk quota purchased for 33.3p per litre of 5p per litre, ie the same as the leasing price.

1989/90 saw a marked change in the balance between the leasing cost on the one hand and interest charges on the other. Leasing prices increased to figures in the region of 6.5p per litre and at the same time, interest charges started to drift downwards.

Many forward-looking farmers (including case history farmers described in Chapter 16) took a decision at this time to purchase milk quota with a view to making savings in the cost of interest compared to the cost of leasing. Banks at that time took a lot of convincing that this was the correct decision. (This in part explains why prices at that time were low and also applies at the present time, 1994, with quota to lease making 8.5p and quota to buy no more than 47p.)

Interest rates continued to drift downwards: In 1990/91 at 12%, the interest on quota purchased at 33p per litre was 4p per litre. At current interest rates (1994) of 7% the interest charge on quota purchased at 33p is only 2.31p per litre.

There is also now the added benefit that milk quota is trading at 47 to 48p per litre giving a substantial capital gain as well as a substantial saving in annual costs. The time to buy or lease quota to get a good deal is when it does not appear to be needed as the national production is or is likely to be under quota. This was certainly the case in 1992/93 when quota could be purchased at less than 30p per litre. A few small lots were purchased by forward-looking farmers at 27 to 28p per litre.

With hindsight, these sales should not have been conducted by the vendors at that time. This illustrates clearly a situation where one dairy farmer's downfall was another farmer's opportunity and illustrates that dairy farming is gradually becoming a tough business in which to operate.

The above is history; it is now necessary to deal with the future and the argument for and against buying milk quota.

The main argument against is that one could be purchasing an asset which one day will lose its value. They are here to stay until the year 2000 but there is no guarantee after that. This is particularly significant for non-milk producers who, in effect, have taken a decision to deal in milk quota, rather than produce milk, as they have sold their cows, and their buildings and equipment are not being updated. These farmers certainly need to be aware that one day the value of their milk quota could disappear.

This is also true in respect of retired tenant farmers, but in this case, their potential loss only relates to the income from leasing, not a capital loss unless they manage to do a deal with their landlord for their share of the capital value. Tenant farmers have always found it difficult to retire and find adequate funds to purchase a house and live comfortably in retirement. The leasing of milk quota has provided a means whereby they may retire from milk production, lease out their milk quota and continue working on the farm with a less profitable enterprise, eg youngstock rearing or beef production. These farmers need to have a contingency plan if quotas should disappear in the year 2000.

The loss in the value of the milk quota, however, is of less significance in relation to a balanced dairy farm where all the quota purchased is being produced, in other words, where there are enough cows on the farm, and appropriate buildings, to produce the quota. If quotas disappeared most of the value of the milk quota is likely to be reflected in the value of the farm as people wishing to enter milk production will have no option but to bid against each other to buy these attractive dairy units. They will also have to bid against each other to buy the required dairy cows. To stand the story on its head, this could in fact be the very time when the efficient dairy producer should take a decision to retire and sell his farm and dairy cows to a new entrant wanting to commence milk production.

The really 'clever' farmer will guess when quotas are going to disappear, get it right, sell his cows at the same time as his quota and rear youngstock to start milk production again when the quota disappears. Or will he? It's more likely he will be made to sell his quota and his cows because of financial pressures and get it right by accident.

TAX CONSIDERATIONS

This has already been touched upon in relation to the purchase of clean quota to overcome a superlevy liability. It is important to realise that milk quota is a capital asset and is treated as such in the tax accounts. Milk quota leasing is an allowable expense against profits in the year in which it is incurred. The purchase price of milk quota is not allowable but the interest on the capital borrowed is allowable as an expense.

A second most important factor is that milk quota purchases are pooled with the existing milk quota for CGT (capital gains tax) purposes. This means that if 100,000 litres is purchased for £50,000 to add to a quota of 900,000 litres, awarded in 1984 at no cost, the base value for all the quota on the farm for future transactions is £50,000, ie 5p per litre.

If 100,000 litres are subsequently sold for, say, £50,000 the capital gain is £45,000. After deducting the annual allowance of £5,800, the tax liability is £9,800 (£39,200 at 25%) if the taxable rate is 25%, £15,680 if the taxable rate is 40%.

The above highlights the disadvantages. There are also advantages. Two people in a partnership can sell quota to the value of £11,600 per year and pay no capital gains tax. This can be an important part of the strategy in relation to a farmer and his wife who are reducing the size of their business. They could sell quota to the value of £11,600 in each year over a period of, say, 5 years giving a total sale of £58,000, free of tax.

In most years the price of quota tends to fall when it is fully used and then rise again at the start of the new quota year. It is therefore often prudent, if a farmer wishes to increase his quota, to lease in the appropriate amount, say 100,000 litres, pay the leasing charge of, say, £7,000, claim tax relief on this £7,000 and then buy in the quota fully used at a discount.

Occasionally, opportunities occur to do this to considerable effect if quota can be purchased from a producer who has to sell quota more than 100% used. This quota could be purchased at a substantial discount, but to take advantage of it the purchaser has to be under quota as the over quota liability is transferred from the vendor to the purchaser.

The new regulations introduced in 1994/95 could lead to yet another opportunity for the shrewd businessman/dairy farmer who has leased in more quota than is required for his own production. He may find that he can sell clean quota at the end of the year at a premium which in turn can then be rolled over into the purchase of more quota in the coming year. Tax relief would be allowed on the leasing in of milk quota but the whole of the gain on selling the clean quota would be reinvested in the quota repurchased. Time will tell whether or not this proves to be true, and if it does then this in turn could lead to yet another revision to the regulations so that this strategy cannot be put into effect.

Returning to the question of capital gains tax and the purchase/sale of quota to avoid superlevy liability at the end of the year. It is a most important principle to sell first and buy second, if capital gains liability on the transaction is to be avoided. In other words, if a farmer cannot afford to retain the extra quota he has to purchase he must first sell his used quota and then buy the clean quota.

TRADING IN MILK QUOTA

The points discussed previously in relation to tax lead to the need to discuss 'trading in milk quota'.

Most dairy farmers are working hard to pay interest on the money they borrowed to buy milk quota, their farm or other assets. The secret in business is to have money working for you. This can best be illustrated by taking a farm that has a base quota of 900,000 litres and an objective to produce one million litres. Such a farmer would normally purchase 100,000 litres and agree an appropriate repayment strategy with the bank, probably to repay the loan over a period of 5 to 6 years with consequent cash flow pressures.

An alternative strategy would be to plan to purchase 200,000 litres, the objective being to produce, say, 100,000 litres and to lease out the 100,000 litres surplus to requirements. At current prices (June 1994) the cost of the 200,000 litres of quota would be in the region of £95,000, which at an interest rate of 7.25% would give an annual charge of just under £7,000. The 100,000 litres surplus to requirements could be leased out at, say, 7p per litre, or £7,000, giving a net additional trading cost to the business of nil.

A few farmers carried out a strategy along these lines two years ago. The income from leasing out their quota has paid the interest on all the quota purchased. The quota was purchased at a price in the region of 30 to 35p per litre and now has a value in the region of 45 to 48p per litre.

TRANSFER OF DIRECT SALES TO WHOLESALE QUOTA

It became fairly common practice in recent years for producers to establish a separate dairy unit on which they intended to produce milk at some date in the future but in the short term only intended to trade in quota. This led to the dubious practice whereby a small quantity of, say, 20,000 litres of Channel Island quota was purchased with a butterfat base of, say, 5.60% and allocated to this holding.

A substantial quantity of direct sales quota was then purchased. Direct sales quota did not have a butterfat base and its purchase price therefore would reflect average quota values, ie 4% quota with a value in the region of 48p per litre. At some stage during the year advantage would be taken of the direct sales/wholesale quota interchange and the whole of this direct sales quota would become wholesale quota at 5.60%, thereby greatly enhancing the value of the milk quota and the amount of butterfat available to the producer.

The loophole in the legislation has now been closed by the announcement from the Ministry that direct sales quota is now presumed to have a butterfat base of 3.80% in the case of the farm where milk is not sold direct. In the case of genuine milk producers/direct sales, the exchange is still at the producer's actual base rate.

CHANGES IN MILK QUOTA PRICES

This chapter is being updated in June 1994, approximately 4 months after it was written. In the text the price of milk quota in February is quoted as being 12p per 1% or 48p per litre at 4% butterfat. This is also the price at which it is trading at the present time, ie in June, but there was a dip in prices in April/May as milk producers who were over quota had to sell quota in order to overcome their superlevy problem.

Case history/example farms are referred to elsewhere in the book and several of these took advantage of the dip in milk quota prices to purchase quota at prices in the region of 10.5 to 11p per 1% BF, ie 42 to 44p per litre. Yet another illustration of the need to be ready for opportunities.

CHAPTER 13

Dairy Farming Prospects

VARIATIONS IN PROFITS FROM YEAR TO YEAR

The trend in profitability in dairy farming during the past 12 years is well illustrated by the data shown in Figure 13.1 taken from the Genus Report on costed dairy farms in 1992/93. These profits are adjusted for inflation. They show that there was a period of prosperity for dairy farmers prior to the introduction of quotas in 1984. This was followed by four lean years before the increase in profitability between 1988 and 1990. This was brought to a fairly abrupt end in June 1990 as a result of the BSE outbreak. If BSE had not occurred profits at that time would probably have climbed even higher and the fall in 1991/92 could have been more dramatic.

Figure 13.1 Profit trends adjusted for inflation

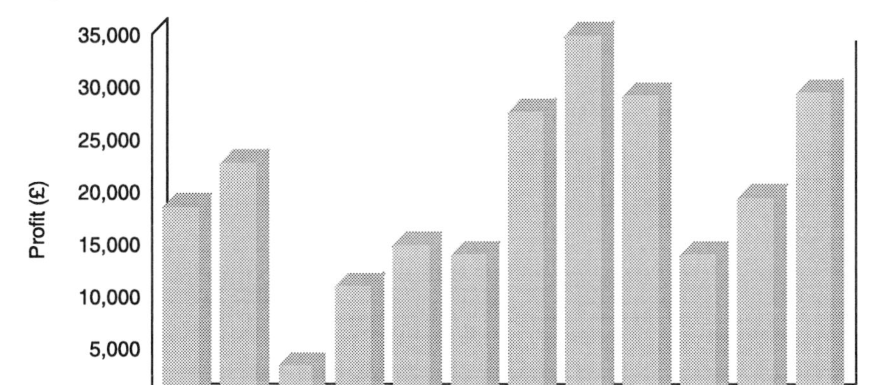

This data shows that the profits in real terms in the year ending 1993 were almost at the level achieved in 1989. Since 1993 there has been a significant increase in the milk price, cull cow and calf prices have been firm, and it is anticipated that the profits shown in the 1993/94 Report will be higher in real terms than those achieved at any time during the previous 12 years, and could be as high as the record profits shown in 1972/73.

As dairy farmers, therefore, we need to be aware that at the present time (1994) we are probably approaching a peak in the profitability of dairy farming and this could be followed in due course by a fall. We need to bear this in mind when preparing our future strategy and study what information we can to give us an idea of future trends.

When preparing budgets and strategic plans we have to make the best estimate we can of the possible future trend in costs and prices. In particular we need to try to estimate what the likely trend is going to be in the number of dairy cows and the potential nationally to produce more or less than quota.

The author has found that the trend in insemination numbers and the number of cows culled from the national herd has given a good indication of the trends that one can expect in dairy cow numbers. These can also be studied, as and when they are released by the government, in the census data taken each year in June and December. It is now proposed to look at each of these in turn and see what indication they give of the possible trends we can expect in the future.

TRENDS IN INSEMINATION NUMBERS

The number of inseminations carried out by Genus give a good indication of the likely trend in the number of calves that can be expected to be born in the future and hence the potential shortfall or surplus of dairy heifers/beef calves from the national dairy herd. Figure 13.2 summarises the number of inseminations carried out by Genus during the six years ended September 1993. There is a substantial fall in the total number of inseminations during these six years from 2.158 million to 1.849 million, a decrease of 14.5% which is in line with the reduction in dairy cow numbers during this period.

During these six years there was relatively little change in the number of inseminations carried out with dairy-type bulls. There was a significant increase in 1989/90 reflecting the profitability of dairy farming at that time and this increase in the number of

Figure 13.2 Insemination numbers

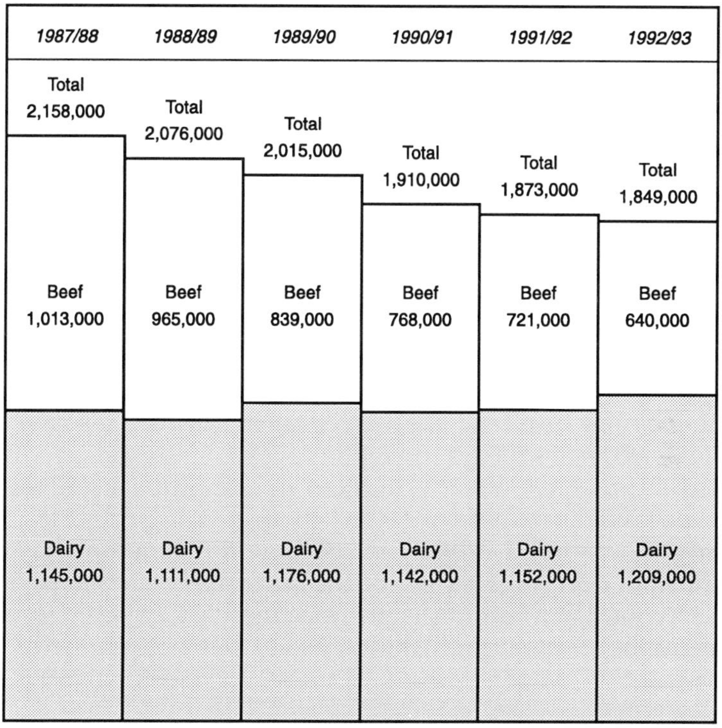

	1987/88	*1988/89*	*1989/90*	*1990/91*	*1991/92*	*1992/93*
Total	2,158,000	2,076,000	2,015,000	1,910,000	1,873,000	1,849,000
Beef	1,013,000	965,000	839,000	768,000	721,000	640,000
Dairy	1,145,000	1,111,000	1,176,000	1,142,000	1,152,000	1,209,000

inseminations in 1989 was a factor the author tried to take into account when predicting the outcome for the milk quota year 1993/94. The increase at that time indicated that there would be more heifers than normal to calve down in the quota year 1993/94, and this has helped to ensure that the country has finished up over quota at the end of the year. However, a factor that could not have been predicted from this data was the increase in imports of dairy heifers from Holland. The milk production obtained from these imported heifers, said to number 20,000, represents a significant part of the amount we were over quota.

There has, however, been a reversal since April 1993 in the trend in the total number of inseminations. From that date there has been an increase month by month in the region of 3% compared to the previous year.*

* Unfortunately, data for the period October 1993 to March 1994 was not available at the time of writing. This data is expected to show a continuation of this trend.

There was also a very significant increase during the 12 months ending September 1994 in the number of cows served to a dairy-type bull. This would suggest there is going to be a surplus of down-calving heifers in the not-too-distant future.

The number of inseminations carried out with dairy-type bulls is likely to continue to increase until the end of 1994 or until such time as there is a down-turn in dairy farm profitability and dairy heifer prices. This means that the number of dairy heifer calves born will probably continue to increase until September 1995 and consequently, the number of heifers coming forward to enter the herd will continue to be at a high level until at least 1997/98, ie 2 to 3 years after their birth.

These insemination numbers also reveal a second, and perhaps more important trend which is the reduction in the number of cows inseminated to a beef bull. These have fallen during the past 6 years from 1,031,000 to 640,000, a reduction of 38%.

These trends have already resulted in a shortfall in calves of a beef type and have been reflected in firm calf prices. It would appear that this trend is likely to continue for some time to come. This shortfall in beef calves *may* be the basis of a continued period of prosperity for dairy farming as the price received for beef calves has a tremendous effect on profitability.

COW SLAUGHTERINGS

The number of cows slaughtered in the United Kingdom is itemised in Table 13.1 for the period January 1988 to April 1994. The data for the two years 1992 and 1993 makes very interesting reading as the number slaughtered each month from January 1992 to December 1993 has always been less than in the same month in the previous year, resulting in a fall in the number of cows slaughtered from 714,000 in 1991 to 570,000 in 1993. This period of low slaughterings has been associated with a period when beef producers were establishing suckler herds (to claim suckler cow premium) and dairy farmers were retaining as many cows as possible as nationally we were under quota. This, coupled with the introduction of more heifers born as a result of the increase in inseminations in 1989/90 and Dutch imports has eventually led to a situation whereby the United Kingdom finished 1993/94 over quota.

The number of cows slaughtered in the period January to April 1994 is 4.6% more than in 1993. This could be the start of a new trend, in other words, an increase in slaughterings associated with a

Table 13.1 Monthly number of cows slaughtered in the UK (thousands)

	1987	*1988*	*1989*	*1990*	*1991*	*1992*	*1993*	*1994*
January		74	76	67	75	70	62	64
February		49	51	50	55	51	48	49
March		46	43	48	50	47	44	43
April		53	59	54	61	54	42	49
May		38	43	37	47	40	34	
June		40	45	34 (BSE)	51	42	38	
July		55	57	56	60	55	53	
August		50	49	41	51	47	42	
September		55	58	53	58	54	47	
October		79	77	71	80	66	64	
November		69	68	64	74	56	54	
December		53	47	51	52	50	42	
Total	898	661	673	626	714	632	570	

Table 13.2 Cow slaughterings as a % of the breeding herd (UK)

	Breeding herd (June) (thousands)	*Slaughterings (thousands)*	*%*
1991			
Dairy herd	2,770		
Beef herd	1,669		
Total breeding herd	4,439	714	16.1
1992			
Dairy herd	2,682		
Beef herd	1,669		
Total breeding herd	4,351	632	14.5
1993			
Dairy herd	2,667		
Beef herd	1,751		
Total breeding herd	4,418	570	12.9
1987			
Dairy herd	3,042		
Beef herd	1,343		
Total breeding herd	4,385	898	20.5

capacity for the dairy industry to produce more than quota, repeating the trend that occurred in 1991 compared to 1990.

Note: Some idea of the capacity of the industry to adjust cow slaughterings to cope with the need for more cows can be judged from Table 13.2. It is often said that the rate of culling from the national herd is in the region of 20 to 25%. The data in Table 13.2 shows, however, that the national culling rate including cullings from the beef herd was only 12.9% in 1993. This is in sharp contrast to 1987. In that year the number of cows slaughtered was 20.5% of the national herd reflecting a need to reduce cow numbers following a cut in milk quota.

DAIRY COWS/HEIFER NUMBERS

The cenus data produced by the MAFF in June and December also provides background information to predict future trends, although many farmers question the accuracy of this data.

Table 13.3 summarises the data for England and the United Kingdom for the three years June 1991 to 1993. Dairy cows in England, plus in-calf heifer numbers in 1992 were down by 2.6% compared to 1991. There was a very modest increase in 1993 (0.09%). This has eventually been associated with the ability to produce over quota in 1993/94 as cow slaughterings continued at a very low level. They were 26,000 less in the 6 months July to December 1993, or down by 7.9% compared to 1992.

Since this section was first written census data (published 30 March 1994) has become available for December 1993. This census data shows a 1.4% increase in the number of cows in the national herd in December 1993, compared to 1992, substantiating the tentative conclusions (drawn by the author from a study of the number of cows slaughtered and the number of inseminations carried out), that cow numbers would increase in 1993 compared to 1992 and that this would result in the country being over quota.

The 1993/94 quota year is history in June 1994. What is likely to happen in 1994/95?

The December 1993 Returns show a 1.4% increase in cow numbers and insemination numbers are continuing to increase which would suggest that cow numbers will be more than adequate to produce quota. On the other hand, dairy inseminations fell in 1990/91 and this was reflected in a 4.3% reduction in the number of dairy replacements in England aged 1 to 2 years in June 1993 (see Table 13.3). In turn, this has been reflected in a 3.3% reduction in the number of in-calf dairy heifers in the December 1993 Returns and

Table 13.3 June census data

	1991	*1992*	*% change 1992/1991*	*1993*	*% change 1993/1992*
England					
Dairy cows	1,938	1,873	–3.35	1,863	–0.53
In-calf heifers	390	394	+1.02	406	+3.04
	2,328	2,267	–2.62	2,269	+0.09
Dairy replacements (aged 1–2 years)	295	304	+3.05	291	–4.27
Beef/dairy heifers male (aged 6–12 months)	569	539	–5.27	523	–2.97
Beef/dairy heifers female (aged 0–6 months)	419	411	–1.91	423	+2.92
	988	950	–3.85	946	–0.04
United Kingdom					
Dairy cows	2,770	2,682	–3.18	2,667	–0.56
In-calf heifers	534	546	+2.25	568	+4.03
	3,304	3,228	–2.30	3,235	+0.22
Beef and dairy replacements* (aged 1–2 years)	645	688	+6.67	623	–9.45
Beef/dairy heifers (aged 6–12 months)	896	857	–4.35	838	–2.22
Beef/dairy heifers (aged 0–6 months)	831	821	–1.20	838	–2.07
	1,727	1,678	+2.84	1,676	–0.12

* Not collected separately for the whole of the UK.

this offsets the increase in the number of cows. The December 1993 Returns for England show a total for dairy cows, plus in-calf heifers of 2,401 thousand compared to 2,395 in 1992, an increase of only 6,000 or 0.2%.

The increase for the whole of the UK is 24,000 or 0.75%. Cow slaughterings for the first 4 months of 1994 are up by 9,000 so it would seem fair to say that the number of cows in the national herd is at an equilibrium in relation to quota. This is reflected in the current firm market prices for newly-calved heifers and the continued importation of dairy heifers from Holland.

The momentum, however, is building up; dairy inseminations

continue to increase; dairy replacements 1 to 2 years of age in England in the December 1993 Returns are up by 0.2%, whereas in 1992 they were down 2.8% compared to 1991; and the number of females under one year of age is up in 1993 by 5.4%, whereas in 1992 it was down by 1.3%.

The conclusion drawn is that the number of dairy cows/dairy heifers in 1994/95 surplus to requirements will be small but adequate to ensure we reach and exceed quota. The effect of this small surplus on dairy cow prices is likely to be modest in 1994/95, but a substantial surplus and consequent fall in prices seem likely in 1995/96 and 1996/97.

MILK PRODUCTION POTENTIAL COMPARED TO QUOTA

The actual milk production potential compared to the milk quota available has a very significant effect on the value of dairy cows as well as the profitability of individual farms. During the past two years, 1992 to 1994, cow prices have risen by at least 50%, ie from £800 to £1,200, reflecting the shortfall in cow numbers relative to quota, and at the present time (June 1994) are still firm even though we finished just over quota in 1993/94.

For a long time, however, there was some doubt as to whether or not milk production would reach quota in 1993/94, and if this had been the case it would have been the third year in succession that it had not been reached. The data examined for insemination numbers etc., however, would suggest that this era is coming to an end and this conclusion is supported by a study of the actual milk production compared to quota on a week-by-week basis for 1993/94, as shown in Figure 13.3.

Figure 13.3 shows the actual sales to the MMB for England and Wales in 1993/94 compared to quota and the actual sales in the previous year. From July 1993 sales to the MMB were approximately 12 million litres per week more than in the previous year and this would have continued right until the end of the year but for the need for producers to cut back in March to avoid paying superlevy.

This has important implications for the future. If this trend continues there will be a potential to produce 300 to 400 million litres more than quota in 1994/95 but of course this will not happen as producers will have to cut back to avoid paying superlevy. In view of this there would seem to be a good chance that dairy cows/heifer prices will be less firm in the autumn of 1994/spring of 1995 than they were in the previous year, but time will tell.

Figure 13.3 Estimated weekly wholesale output and quota (adjusted for butterfat)

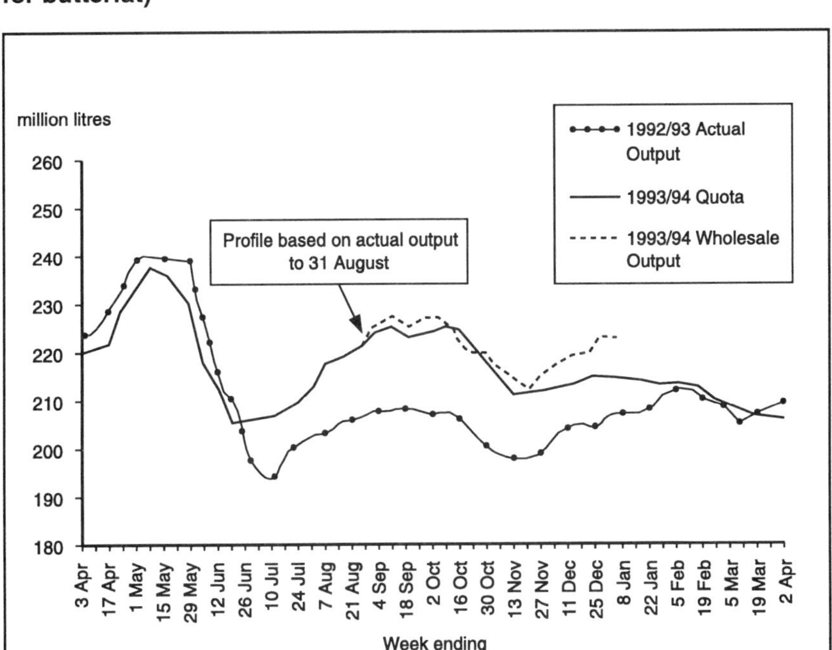

The writer is very conscious that predicting the future is always dangerous. The study of these trends in the past, however, has proved to be profitable particularly in relation to the expected trend in dairy cow prices and the expected trend in the price of milk quota, both to buy and to lease.

An important part of the strategy on farms with which the author is associated has been to have surplus cows/heifers available for sale in 1993/94 and many of these have been sold at a profit, some on the herd basis. The opportunity was also taken in 1992/93 to purchase milk quota and it is now comforting to note that its value is 50% more than its purchase price.

As mentioned earlier, there are now a substantial number of dairy heifers in the pipeline, many of which have still to be born and will continue to be available to enter dairy herds until 1997/98. It is considered, therefore, that during the next three to four years we will have the potential to be well above quota and this potential for over-supply could be exacerbated by a milk quota cut if a cut was imposed. (This, however, now seems less likely than it was, say, two years ago as the European Commission could decide not to impose quotas so that European milk prices fall close to world prices in accord with the GATT Agreement.)

GATT AGREEMENT

The implications for the dairy sector of the GATT proposals are difficult to interpret. A substantial proportion of the 20% reduction which has to be achieved in internal support has already been met.

The reduction in the volume of subsidised exports could have an adverse effect on EC dairy farmers including those in the UK, as could the increase in imports, particularly increased imports of Cheddar and New Zealand butter.

The UK is the largest producer of Cheddar in the EC and the UK is particularly vulnerable to increased Cheddar imports. The GATT proposals could also involve up to a 3% cut in milk quota by the year 2000. The CAP reform plans already agreed include a 2% cut in quota during the period 1994/96 but, as mentioned above, there is some doubt now whether or not these will be implemented. On balance it would seem reasonable to assume that although the GATT proposals will have an adverse effect on dairy farming profitability, these will not be penal and are unlikely to affect significantly the price for liquid milk and other high valued products which are the main strength of the UK market.

MILK PRICE RELATIVE TO FEED PRICE

Indirectly, the GATT proposals should help the British dairy farmer as they will result in a reduction in the price of feed.

The vesting date for the end of the Milk Marketing Board has now been fixed as 1 November 1994, based on revised proposals submitted by the MMB to the Ministry.

The 'freeing up' of the milk market has already been reflected in a significant improvement in the milk price as purchasers compete for the shortfall in supplies in the era following the demise of the MMB. This, together with the reduction in feed costs referred to earlier, should result in an even better milk-to-feed-cost ratio, at least in the short term.

The potential for an increase in home-produced beef production is contained by the suckler cow scheme. Milk quotas will impose a constraint on any substantial increase in cow numbers and these two factors will tend to lead to firm calf prices. One is therefore tempted to draw the conclusion, rather dangerous but hopefully correct, that dairy farmers are set to enjoy a relatively stable period of prosperity during the rest of the 1990s. The author, however, is very well aware that in the past, periods of prosperity in dairy farming have always been rapidly followed by periods of difficulty, ie 1972/73 followed by 1974/75; 1977/78 followed by 1979/80; 1982/83 followed by 1984/85; and 1988/89 followed by 1991/92. Hopefully this will not be the case in the rest of the 1990s.

MILK QUOTA LEASING PRICES

The extent to which an individual farmer enjoys this period of relative prosperity will depend upon the size of his quota relative to the needs of his business.

During the period 1992 to 1994 it was possible to lease in quota at prices in the region of 5p or less. Typical gross margins achieved per litre during this period were in the region of 16p, for example, a cow giving 6,000 litres would produce a gross margin in the region of £960. After deducting the cost of leasing there was a net gross margin in the region of 11 to 12p per litre.

The 1994/95 quota leasing season has started off with very firm prices in the region of 7.5p per litre and there is talk that these could reach 10p per litre before the end of the year. In other words, farmers who need to lease in quota are likely to find that half the gross margin, or more, is swallowed up by the leasing charges.

This trend is likely to continue for the next 2 to 3 years, and unlike the situation 5 or 6 years ago the value of the cows on the farm is likely to go down rather than up. Dairy farmers with limited milk quota are therefore at another crossroads. They need to look carefully at whether or not they should continue to lease in quota.

An alternative strategy for the next 2 to 3 years could be to reduce cow numbers in line with the quota available and introduce or expand other enterprises on the farm, eg arable enterprises, dairy heifer rearing, beef cattle rearing or in certain instances, introduce bull beef, or other non-land-using enterprises as outlined in Chapter 7. These decisions will need to be made in good time, ie before other farmers have realised the necessity to make similar changes.

These alternative strategies will not be significantly more profitable than continuing to produce more milk and lease quota, and they could be less! The brave therefore will decide to buy quota, make good profits after deducting the interest charges involved, and accept the risk that it may lose its value in the knowledge that this would be offset to an appreciable extent by the enhanced value of their farm and dairy herd.

CHAPTER 14

Taxation and the Dairy Farmer

MAKE IT, KEEP IT!

Having spent some considerable time preparing strategic plans, working hard, monitoring performance and achieving good profits before tax, the next management objective is to aim to keep as much of this profit as possible by making legitimate provisions for tax.

There are various steps that can be taken to mitigate the potential tax liability and often these steps need to be taken before the end of the accounting year. Most farmers do not discuss their accounts with their accountants until well after the end of the accounting year, and thereby lose opportunities to save tax.

Accountants deal with all types of businesses and it is sometimes difficult for them to keep up to date with agricultural problems. Farmers are usually most anxious not to pay tax, but often set out to save tax in the wrong way, for example by spending money on machinery and equipment that they do not really need.

A tax revolution, however, is on the way, and this is going to have an impact on all dairy farmers who trade either as sole traders, or in a partnership.

DO-IT-YOURSELF TAX FOR THE SELF-EMPLOYED

The Chancellor announced a new system in his March 1993 budget, which will come into operation from 5 April 1997. Self-employed taxpayers, as from that date, will become responsible for calculating their own tax liability as well as completing their tax returns. These returns will have to be filed with the Inland Revenue by 31 January following the end of the tax year.

The tax will be payable in three instalments, not two, as is the case with the present system. Tax will also be payable on the 'current year basis' (CYB), not 'previous year basis' (PYB), as at present.

At present, a self-employed dairy farmer or farmer in a partnership pays tax in two instalments, ie in January and July, for the profits made in the previous tax year. In other words, the profits made in the tax year ending 5 April 1994 are paid in January and July 1995. If a farmer has an accounting year which ends on, say, 30 April then the accounts for the year ending 30 April 1993 are used as a basis for paying the tax in 1995. A farmer who has accounts ending 31 March 1994 also pays tax in January and July 1995.

Tax payments in the future will be paid in three instalments. An interim payment will be made on 31 January in the tax year and a similar interim payment on 31 July (each instalment being half of the previous year's tax bill). A final balancing payment/repayment claim will be made on the following 31 January.

These concepts are difficult to grasp, especially the transition to tax being paid on the 'current year' basis. 1996/97 will be a transitional year for this switch from PYB to CYB and will entail averaging two years' profit. In the case of a farmer with a tax year ending April, the two years ending 30 April 1996 will be averaged and this will form the basis for the tax payable on 31 January and 31 July 1997 with the final balancing payment being made in January 1998. In the case of a farm with a March year end, the two years ending 31 March 1977 will be averaged to determine the tax payable in January and July 1997 with the final balancing payment being made in January 1998.

There is scope for exploiting this transitional year to the tax payer's advantage, and in some cases it could be beneficial to change the year-end date. This topic is obviously of tremendous interest at the present time and much more could be said. Each individual farmer's tax situation, however, is different, and they are advised to contact their accountant without delay to discuss the implications of these proposed changes.

STOCKTAKING VALUATIONS FOR INCOME TAX PURPOSES

Valuations prepared for income tax purposes are prepared on a quite different basis from those for management accounts. The valuations in management accounts are prepared based on the potential sale value of the animal but in valuations prepared for income tax purposes, the principle is that 'stock falls to be included at the lower of cost, or net realisable value'. This means that farmers in effect need two sets of accounts, one the management accounts to show how much profit is actually being made based on potential sale values and the other to show the profit for tax purposes based on costs of production. In effect, one set of books are required but two sets of valuations. The cash flows should reconcile in both sets of accounts, and so should the revenue and expenditure but the profits shown can be quite different, due to the different treatment of the valuations.

It is quite clear that the stock valuation is a very important aspect in the preparation of both management and income tax accounts. In management accounts one would expect to see a detailed breakdown of the management valuation and how it is arrived at, but the information given in the income tax accounts is usually just one line, ie stock valuation as prepared by the farmer, or as prepared by A N Other agricultural valuer.

The Inland Revenue have recently made a comprehensive announcement on farmers' stocktaking valuations. This appeared in 'Business Economic Note 19' on 31 March 1993. This announcement included the following statements:

> '. . . for tax purposes, we are looking for a figure (commonly referred to as a valuation) which represents the costs, or if lower, the net realisable value of stock'.
>
> 'Valuations problems can be complex and farmers normally seek the assistance of accountants and agricultural valuers and surveyors, but this is not compulsory and some farmers prepare their own valuations.'

Agricultural valuers and surveyors are usually not fully aware of the actual costs of producing animals and crops, when they prepare valuations, as the farmers do not have adequate records. It has therefore become common practice to use 'deemed cost' as the method of valuation.

DEEMED COST

What is deemed cost? To put it simply, it is a percentage of the open market value of the livestock or produce on the date of the valuation. The acceptable percentages are 60% for cattle, 75% for sheep and pigs, and 75% for harvested crop produce.

What are the advantages and disadvantages of using deemed costs? The main advantage is that it is a very simple way of estimating the costs of production compared to making estimates based on the expenditure incurred during the year.

There are two main disadvantages. First, market prices go up and down and can fluctuate widely from year to year. Second, and possibly more important, livestock values have increased very considerably during the past two years, but the actual costs of production have tended to fall. Valuing livestock on a deemed cost basis, therefore, will lead to a higher tax payment than if the valuation was based on actual costs of production.

Many dairy farmers now keep management accounts as well as producing audited accounts for tax purposes. These management accounts allow detailed estimates to be made of the actual costs of production and there would now seem to be a good case for farmers starting to prepare their own valuations for tax purposes, based on their management accounts.

(Note: The deemed costs *must not* be used for purchased animals if they are less than the original price of the animal plus the cost of rearing it from the date of purchase to the valuation date. Deemed cost valuations, therefore, are only valid for home-bred or home-reared stock.)

PRODUCTION COST

Providing the information on costs can be obtained, it is now more appropriate for dairy farmers to produce their valuations based on the production costs, not deemed cost. Business Economic Note 19 lists the items that need to be included in these costs as detailed below.

1. Crop produce and growing crops

The valuation of these should include the cost of the following:

Seeds, fertilisers, and beneficial sprays (sprays used to remedy a particular infestation or crop deficiency do not need to be included)
Grazing lets (normal farm rents do not need to be included)
Drying and storage
Paid and contract labour/machinery costs incurred on cultivations, crop work and harvesting (the notional cost of the farmer and his partner does not need to be included).

Information on the above should be available on most farms with well-kept management accounts and these costs are likely to be significantly less than those based on the deemed cost method of valuation. The cost, for example, of growing a crop of wheat is likely to be in the region of £180 per acre, £60 per tonne for a 3 tonne crop, £50 if the yield is 3.6 tonne. The deemed cost would be in the region of £75 per tonne, 75% of £100.

2. Livestock

The stock valuation should include the following:

Actual purchase price of stock purchased
Insemination costs plus maternal feed costs in excess of maintenance required to produce a calf (not its sale value), plus the cost of rearing to the valuation date, including: feed costs (including forage), vet fees (including drugs), drenches and other medicines, ringing, cutting and dehorning
Paid labour and contract labour costs (The notional cost of own labour for proprietors and partners does not need to be included)

The cost of an in-calf heifer based on the above is likely to be in the region of £400. The deemed cost is likely to be in the region of £600, 60% of £1,000.

3. Livestock numbers

This is not directly concerned with the matter of valuation, but it cannot be stressed too strongly that the numbers shown in the accounts should reconcile with the numbers in opening valuation and the number of calves born during the year.

Management accounts are again an aid to preparing this information as they record the transfer of animals from the dairy herd into the youngstock to calculate enterprise gross margins.

4. Overhead or indirect costs

The inclusion of these costs in the valuation is optional and they do not have to be included except where they have been included in the past and to omit them would be inconsistent.

Examples of indirect costs are:

Depreciation and maintenance of farm buildings
Rent and rates
General labour and machinery costs

In other words, the costs that have to be included are the *direct costs of production*. These direct costs are tending to go down not up and reinforce the need for farmers to think in terms of preparing valuations based on cost of production, not deemed costs.

5. Residual manurial values

These are often an important part of the ingoing on a tenanted farm. They can be ignored for the purposes of the annual stocktaking valuations for tax purposes, as the Inland Revenue accepts that fertiliser applied can be regarded as being exhausted, following harvest.

6. Consumable stores

It is usually relatively simple to value seeds, sprays, feedingstuffs and other purchased consumables based on their actual cost of purchase, providing one has access to the appropriate invoices and a note of the quantity on hand at the end of the tax year.

HERD BASIS

The principle of the herd basis is that the dairy cows enter the herd based on their cost of production, as itemised above, and the animals remain in the herd at this cost, even if there is a very substantial increase in the value of dairy cows.

Farmers who have dairy herds that have been established for, say, 25 years will have cows on the herd basis at figures of no more than £100 per head, ie the cost of rearing the animals when the herd was first formed.

It is difficult to overstate the value of being on the herd basis from a tax-saving point of view, providing that its principles are well understood and it is operated effectively.

If there is an increase in the herd size during a particular year, the additional animals are valued at either their purchase price, or their cost of rearing. It is therefore important to manage the herd in such a way that the increase in the herd takes the form of additional home-reared heifers, not purchased replacements, as the former will be able to enter the herd at a value in the region of £400 to £450, whereas the latter would enter the herd at, say, £1,200.

An important principle regarding the herd basis is that the cow replacing the animal that has to be sold should not be intrinsically better than the cow sold. It is quite permissible, for example, to purchase a cow for, say, £1,200 to replace an aged cull that might sell for only £350, as the intrinsic value of the animal is the same. It is, however, not permissible to purchase a pedigree cow for, say, 5,000 guineas and transfer this into the herd as a replacement for an ordinary commercial cow. Home-reared pedigree heifers can be introduced into the herd, however, as their cost will reflect the initial cost of insemination fees, plus the maternal costs involved in producing the calf, or alternatively the actual purchase price of the animal when it was a calf.

The herd basis system works well for the established dairy farmer, but there are difficulties to overcome for the new entrant. If he/she establishes a herd of, say, 40 cows with an average purchase price of £1,200 this will be the figure at which they will have to enter the herd at the end of the year. The actual value at the end of the year, however, could be considerably less than their purchase price. In the initial years, a new entrant to dairy farming will find it more tax effective to treat the dairy cows on the trading basis and to value them at the end of the year, based on their 'net realisable value', which would probably be mid-way between their purchase price and their cull value. In the long run, it will probably pay the new entrant to be on the herd basis, not the trading basis but this could make life very difficult in the first few years. The difficulties occur as the business has to make a decision at the outset whether the herd is to be treated on the herd basis or the trading basis and once this decision has been made, it cannot be changed until such time as there is a fundamental change in the business organisation, eg a partnership becomes a sole trader, or a sole trader takes in a partner and starts to farm on a partnership basis.

One way to overcome this problem is to start off as a sole trader and then at a later stage form a partnership. The cows would be transferred into this partnership based on their net realisable value.

The main reason for going on the herd basis is that when the herd comes to be sold the profit on the sale is not subject to tax. The rules regarding these sales are very precise; there is no problem if a complete herd dispersal takes place, but there can be problems if a short-term reduction is made.

The Inland Revenue have granted a concession whereby no tax is payable on the profit made on the sale of part of the herd providing this is not less than 20% of the herd, and the numbers are not increased to the original number within a period of five years.

Reduction sales can prove a good means of providing funds for investment in capital assets or reducing borrowing (see Chapter 15). They can, however, have an adverse effect on the cash flow, unless the animals sold are truly surplus to requirements and need considerable long-term planning. Now (1994) is a good time to be having a herd reduction sale as cow values are at a record high.

The important point, as already mentioned, is to have 20% surplus animals available for sale. This is only likely to be the case if plans were prepared two to three years ago (ie in 1991/92) to have a sale at this time. The author has been working closely with several farmers who, during recent years, have had exactly this strategy in mind. The plans in certain instances were prepared four to five, not two, years ago. In other words, it has taken rather longer than expected for the boom times to arrive. Having arrived, however, they are much better than anticipated.

TAX FACTS

The most salient tax facts are set out in Appendix 2 for the year 1994/95.

A few observations are worth making at this stage about these tax facts.

1. Each person has a personal allowance of £3,445 and can earn this sum, ie £66.25 per week or £287 per month, without paying any income tax. However, the amount that can be earned without becoming liable to employees national insurance is only £57 per week.

   If an employee is paid £66 per week, this is not subject to income tax but the charge for national insurance will be 2% on the first £57 and 10% on the amount between £57 and £66 per week. This result is a total payment for the extra £9 per week of approximately £2, or approximately 20% of earnings, a similar amount to the tax paid on the income between £3,445 and £6,445 per year. As a result, many people are employed on a part-time basis, based on a wage of not more than £57 per week.
2. Attention is drawn to the fact that the capital allowance on agricultural buildings is only 4% per annum and this is on a straight line basis. The capital allowances on machinery and plant are at 25%. This means that it is relatively much more difficult to fund agricultural buildings out of cash flow than it is plant and equipment, as the investment has to be funded out of taxed income with very little tax relief. It is therefore desirable, for tax reasons as well as pride, to aim to construct most of the internal fittings to a building with farm labour as opposed to using building contractors.
3. An individual can earn approximately £27,245 before becoming subject to tax at 40%, which means that a partnership of a husband and wife can earn £54,490 before tax is levied at this higher rate.

Partnerships are a good means whereby the profit payable in a farm business can be contained. However, if the farmer's wife is shown in the books as a partner, it is essential that she is involved with the day-to-day management of the business by carrying out duties such as assisting with the bookkeeping and being involved with meetings with people such as the bank manager and the accountant, even if she is not involved in the day-to-day manual work. Sons can also be brought into the partnership and this too can help to reduce the potential income tax payable in successful businesses.

In this connection, however, it should be noted that certain tax allowances for expenses can be incurred in a business for an employee, but not for a partner. Repairs, for example, can be carried out to a farm cottage for an employee and these can be treated in the accounts as property repairs and completely written off against tax. If the farmer's son or daughter is an employee, this is an allowable expense, but if he or she is a partner it is regarded as an expense incurred on his or her private residence and is not an allowable expense. Improve the cottage, therefore, before making your son or daughter a partner in the business.

4. Returning to partnerships, the division of profits between partners can be varied from year to year. This can allow flexibility in the gradual transfer of assets from the older generation to the younger generation, as the younger partners can receive a higher proportion of the profit, always assuming of course that the older generation still has 'enough to live on', and is still not doing most of the work!

CAPITAL GAINS TAX

Capital gains tax is levied as the 'top slice of income' but there is an annual exemption per person of £5,800. Milk quota sales are subject to capital gains tax. It is not possible to buy, say, 50,000 litres and then sell the same 50,000 litres and not pay tax, as the 50,000 litres purchased is added to the 'pool' awarded to each producer in 1984.

If we assume that the producer was originally awarded a quota of 200,000 litres, and that the 50,000 litres cost £25,000, it would be deemed that the total cost of the quota was £25,000, divided by 250,000 litres, which equals an average of 10p per litre. If 50,000 litres are subsequently sold, the Inland Revenue would say that a capital gain had been made of 50 minus 10, or 40p per litre. This would give a capital gain of £20,000. This would be subject to tax, less the individual allowances of £5,800 per year.

This takes one on to consider 'roll over'. The capital gain made on the sale of milk quota can be rolled over into another asset, providing that this investment is made within a period of three years of the date the milk quota was sold. The tax on the capital gain, however, will become payable in the December following the tax year in which the gain was made. For example, let us assume that milk quota is sold on 31 July 1994. The tax due on this gain will become payable in December 1995, but this can be reclaimed if the whole of the proceeds are re-invested in an eligible investment before 31 July 1997.

TAX PLANNING

This chapter has covered very briefly some of the points to consider in relation to both income tax and capital gains tax. This is by no means meant to be an exhaustive study of the subject. Inheritance tax has not been mentioned as this is a subject in its own right. Each individual person, whether he be a farmer, dairy herds manager or herdsperson, needs to study the tax facts shown in Appendix 2, examine some of the ideas mentioned in this chapter, and discuss these with his accountant or local tax inspector.

Only recently, ie since the author started writing this book, he has been in conversation with a farmer who has just handed over his dairy farming business to his daughter. He was full of praise for his accountant who had persuaded him each year to reinvest some of his profits in a pension fund, ie not all back in the farm. This led to significant tax savings at the time but, perhaps more important, means he now has an income from outside the farm and this is a major reason why he has been able to hand over his business to his daughter.

The chapter began with the statement that having made the profit we should aim to keep as much of it as possible. The message to end this chapter is 'think tax savings'.

CHAPTER 15

Capital Requirements and Borrowing Money

CLASSIFICATION OF FARMING CAPITAL

The word 'capital' has a precise meaning to economists and is regarded as a factor of production separate from land. It is also kept distinct from money, which is simply a means of exchange and a symbol of capital. A dairy farmer's capital is usually expressed in pounds sterling, but the actual farming capital is the land, milk quota, buildings, livestock and other assets in the dairy farming business. In this chapter, the word 'capital' is used in the popular sense, ie as a measurement of the money required to buy or rent a farm and to farm it effectively.

The capital invested in agriculture is often classified according to the systems of land tenure, ie into 'landlord capital' and 'tenant's capital'. This served a very useful purpose when the vast majority of land was farmed on a tenanted basis. Today, however, this is not the case because over 70% of land is owner-occupied and many farms have a mixed assets base, ie part owned, part tenanted.

It is also usual to divide capital into two classes according to whether it is 'fixed' or 'working (circulating)' capital. Fixed capital concerns items that do not wear out quickly and are replaced infrequently, such as buildings and equipment. Working or circulating capital is that which is used or consumed in a single process and is replaced once per annum or more frequently, eg money required to pay wages and purchase seeds and fertilisers. This simple division, however, is not convenient in farming as it is difficult to know where to place items such as dairy cows and other livestock.

We need to bear the above classifications in mind but in farming it is better to adopt a classification based on the nature of the goods and services and whether one would regard the capital investment as long-term, medium-term or short-term, as shown below.

1. Long-term capital investments
   Land
   Milk quota
   Buildings
   Other fixed equipment
2. Medium-term capital investments
   Breeding livestock
   Machinery and equipment
3. Short-term capital investments
   Non-breeding livestock
   Seasonal working capital, including seeds, fertiliser and cost of cultivations.

Milk quota is shown as a long-term asset as most dairy farms would not purchase it with a view to sale. Milk quota, however, unlike land, is a very tradeable asset and could therefore be regarded as a medium-term asset to be sold if and when the price is right.

The term 'tenant right' is one with which tenant farmers are familiar, particularly those seeking the tenancy of a farm for the first time. This is the sum of money that an ingoing tenant has to pay to the outgoing tenant. It often includes medium-term and long-term investments such as milking parlour equipment and tenant's fixtures but most of the tenant right is likely to be tied up in short-term investments such as the value of crop produce on hand, crops in the ground and unused manurial values.

MILK QUOTA

The capital value of milk quota on dairy farms is, to say the least, very substantial. At the present time (1994) it is worth in the region of 47p per litre.

Prior to the introduction of quotas the major investment on an owner-occupier dairy farm was the land. The land can be separated into two component parts, in effect the bare land and the milk quota that is attached to this land.

Tenants often purchase milk quota having obtained the appropriate agreements from their landlord and this makes a very substantial part of their assets required to farm.

Farmers' balance sheets prepared for tax purposes show the

value of the milk quota that has been purchased but do not, of course, include the value of the quota that was awarded to them in 1984 as this had a nil cost.

To illustrate the significance of milk quota let us look at the price that a well-equipped dairy farm of say 200 to 220 acres (80 to 90 hectares) would make at auction in 1994. The farm would probably be valued by the auctioneer for the prospective vendor along the lines shown below:

Land and buildings	£400,000 (£1,800–£2,000 per acre)
Milk quota	£360,000 (say 800,000 litres at 45p per litre)
Farmhouse	£180,000
	£940,000

How much the farm would actually make at auction would depend upon the number of bidders. Providing there were at least two farmers keen to buy the farm, and two bankers who were keen to support the prospective purchaser, the farm could reach a price in the region of £1 million.

If the farm was less well equipped and less endowed with milk quota, say 400,000 litres instead of 800,000 litres, then the farm might sell at auction for a price in the region of £750–£800,000. The prospective purchaser would probably be a non-dairy farmer and he would subsequently sell off the milk quota for, say, £180,000, leaving him with a youngstock farm at a cost in the region of £600,000.

This demonstrates the significance today of the value of milk quota when selling a dairy farm. At this stage it is also important to emphasise the point mentioned above, that there must be two keen purchasers and two bankers prepared to help fund the purchase of a farm. The willingness, or otherwise, of bankers to lend to farmers to buy farms has an undoubted effect on the eventual sale price.

In 1993, bankers were rather unwilling to lend capital to farmers to buy farms, partly because the banks had experienced two years of difficulty and were not generally aware of the improvements in profitability that were taking place in dairy farming. Mention has been made that the farm quoted above could make £1 million in 1994. In early 1993 it would probably have had difficulty in reaching £800,000, and with hindsight this was a good time to try to buy a farm. The previous chapter dealt with the need to study trends and prices with a view to being in a position to take opportunities, if and when they arise. In the spring of 1993 the author worked closely with two farmers who had some difficulty in persuading their bankers to give the necessary support, but they eventually obtained the necessary capital and did purchase farms. It is now a great pleasure to look back and see that these were a timely decision.

CAPITAL REQUIRED TO FARM AS A TENANT

As mentioned previously, very few people today get the opportunity to start farming as a tenant, but opportunities do occur from time to time, particularly on County Council holdings. On a landed estate it is also normal practice to separate the business of farming from the business of owning land. A notional rent is charged for the land in hand, so calculations can then be made of the working capital required to run the farm and the return on capital that can be expected/achieved.

Table 9.2 (Chapter 9) itemises the capital invested in a typical 80 hectare farm. The investment, excluding the 20 hectares of land owned, is put at £206,500, in other words just over £2,500 per hectare or £1,000 per acre.

These estimates are based on the written-down-value of the machinery and equipment. More importantly, the dairy cows are valued at £700 per head which is considerably less than the price that would be required to purchase them newly-calven. The capital required by a new entrant to farming could be some 20 to 30% more than those shown in this table.

This immediately leads to a problem when one tries to talk in farming about such things as the return on capital: Should the dairy cows be valued as shown in the table, at £700 per head, or at the cost the farmer incurred to rear them, say £500, or should they be valued at say £900 per head to reflect their current value?

It is particularly difficult to value dairy cows at the present time (1994) as newly-calven cows are trading at £1,200 per head, whereas less than two years ago they could be purchased for £800 per head. The cull value of a cow, ie £450 to £500 per head, is no more than it was two years ago, in other words when the cow leaves the herd its value has not gone up. Costs of rearing these animals have also not gone up, in fact they are tending to go down. This is commented on in Chapter 14, dealing with taxation, which stresses the importance of aiming to put stock in valuation at their cost of production, rather than at their 'deemed cost' (arrived at by taking a percentage of the market value of the animal).

Finally, when discussing the capital requirements for dairy farming we have to be careful that we are considering the *total* capital requirement, ie the total investment required to run the farm or the net capital investment required after deducting borrowed capital.

The tenant's capital required to operate the farm, outlined in Table 9.2, is £206,500. The actual net worth is £214,000 after deducting the liabilities and adding back the value of the 20 acres of land that is owned. The finance charges resulting from the purchase of this land are in effect paid for by the land that is tenanted and this is the way, over the years, in which today's successful dairy farmers have progressed, starting off as tenants and gradually becoming successful owner-occupiers as well as tenant farmers.

It is often said that today's young farmers have much less opportunities than their forebears, but there are still a few young farmers who are brave enough to take their opportunity. Reference was made earlier to two farmers who managed to purchase farms at the right time in 1993. One of these was a tenant farmer's son. The bank took a bit of convincing that the deal was on, and obviously the farm will not be financed and paid for without some difficulty, but the young person concerned is confident that in the end his desired objective will be achieved. He already has the satisfaction of knowing that its appreciation in value in the first year is well above the interest charges he has paid.

WHAT IS LAND WORTH? WHAT IS MILK QUOTA WORTH?

These questions are often asked and are very difficult to answer: A simple answer is, slightly more than what the under-bidder is prepared to pay.

How much is a dairy cow worth? Is £1,200 too high, or is it about right? In order to answer these questions we need to be able to relate the value of the dairy cow, or the land, or the milk quota to some other asset, or perhaps even to the retail price index.

The wages paid to farmworkers tend to follow the retail price index quite closely, as in most years the Agricultural Wages Board award is made by the independent members and this is based on the retail price index.

In 1971 the average price of a dairy cow was in the region of £120 and the minimum agricultural wage was just under £15 per week, in other words, one cow equalled eight weeks' wages. 1972/73 was a period of excellent profitability in dairy farming and during that period the price of cows rose to approximately £200 per head and this was equivalent to 11 weeks' wages, as the average wage was in the region of £18 per week.

Profitability of dairy farming fell back to more normal levels from 1974 to 1983, and during this period the price of the average cow was equal to approximately 8 weeks' wages. After quotas were introduced in 1984 the price of the average cow fell back to no more than 4 to 5 weeks' wages, and continued to be low in real terms until production nationally fell under quota in 1991/92.

A farmworker's minimum wage at the present time is £145 per week and multiplied by 8, this is just under £1,200. In other words, it could be said dairy cow values relative to wages have now returned to normal levels.

The period between 1984 and 1992 was exceptional in the sense that, during that period, cow prices were relatively low due to the fact that milk production throughout the period was potentially well above quota.

Milk quota recently has been trading at approximately 48p per litre. 100,000 litres has cost in the region of £48,000, and this is equivalent to 40 dairy cows at £1,200 per cow.

Five years ago the average price of a dairy cow was in the region of £600, and the price of milk quota was 35p per litre. At that time 60 cows were required to purchase 100,000 litres of milk quota. Milk quota prices at the present time, relative to cow prices, are lower than they were five years ago. The difficult question to answer is which way will they move in the future? Will the price of cows fall to say £1,000 per head so that 50 are required to purchase 100,000 litres at 50p per litre? Or will the price of milk quota increase to say 60p per litre so that once again it takes 60 cows to purchase 100,000 litres?

The price of milk quota can also be looked at relative to the price of milk. Four to five years ago the average price of milk per litre was less than 50% of the price of milk quota, ie an average milk quota price of, say, 35p per litre, and milk prices ranging from 15p to 17p per litre. Milk quota prices dropped back in 1992 to a figure in the region of 30p per litre, and at that time milk prices were in the region of 19p per litre. Farmers who had the courage purchased substantial quantities of milk quota at that time, in the expectation that their values would rise, and so far they have been proved right.

We now need to turn to look at the price of land, and in this case to discuss it initially in terms of the weekly minimum wage paid to a farmworker.

During the period 1971 to 1983 there was a rapid escalation in the price of land, which increased from £200 per acre (£500 per hectare) to £1,600 per acre (£4,000 per hectare). During this period the price of land per acre, as a multiple of the weekly wage, ranged from 20 to 40 and averaged 25. For example, the minimum agricultural wage for an agricultural worker in 1981 was £64 per week and the price of land at that time was £1,600 per acre (£64 × 25).

The minimum agricultural wage today is approximately £145 per week, which multiplied by 25 equals £3,625 per acre (£9,000 per hectare). This is well above the value today of bare land, but if one also takes the value of milk quota into account, there is a fairly good correlation with the price paid in 1981.

At the start of this chapter, the value of a 200 acre farm was discussed. The land including quota but excluding the value of the farmhouse was put at £760,000, or between £3,378 and £3,800 per acre. This is very similar, in real terms, to the prices being paid in 1981. Questions are often asked as to what would happen if the value of milk quotas suddenly disappeared. The above information would suggest that most of the value of the quota would return to the land.

OPPORTUNISM

Returning to the opportunist farmer who studies trends: He purchased 100,000 litres of milk quota in 1992 for £30,000 when cows were worth £700 to £800. He is now (in 1994) selling 25 cows at £1,200 per cow to pay for this quota.

One theme in this book is opportunism and this illustrates perfectly what has actually happened in respect of several farms with which the author has been closely involved during the past few years.

BORROWING CAPITAL TO FARM

Sources of credit are classified in the same way as the need for capital investments and the major sources are summarised below:

1.	Long-term loans for land purchase and major capital improvements to buildings and fixed equipment	Agricultural Mortgage Corporation Joint Stock Banks Farmer's relatives
2.	Medium-term loans for machinery and equipment, minor capital improvements and major expansion of livestock numbers	Joint Stock Banks Hire purchase companies Machinery and equipment leasing companies
3.	Short-term credit facilities to cover seasonal variation in capital requirements for seeds, fertilisers and livestock	Overdrafts from joint stock banks Merchants' credit Special finance schemes financed by banks

The Agricultural Mortgage Corporation (AMC) grants loans only on the 'mortgage security' of property, so loans from this organisation are available only to owner-occupiers. Loans are granted for periods of up to 40 years but most loans are taken out for periods not exceeding 25 years. A case for borrowing for a period longer than 15 to 25 years is difficult to justify unless one is able to borrow at the start at a low fixed rate of interest and considers that interest rates are not likely to rise. A farmer who borrowed at a *fixed* rate of 7% in the 1960s made considerable savings compared to the actual interest rates charged during the 1970s and 1980s, in a period when interest rates were high, ie 14 to 15 % for long periods of time.

SHOULD WE FIX INTEREST RATES?

Interest rates at the present time (May 1994) are quite low (base rate 5¼%) so many farmers are borrowing at an effective interest rate of 7¼%, ie 2% over base.

Farmers borrowing at 2% over base tend to be well established and can offer significant amounts of security. New entrants to farming and those with less security are being charged rates as high as 5% over base. A feature of meetings with bankers during the past two to three years has been their desire to try to increase the rate over base which is charged to farmers.

At the present time, most of the banks and the AMC are suggesting to farmers that they should 'fix' their interest rate as a hedge against future rises in the interest rate level. These fixed rates, however, are at least 1½ to 2% above the rate that can be obtained on a variable basis. During 1993 the author was involved with several farms where it was necessary to 'unscramble' fixed interest loans that had been set up by the farmers in 1991/92 at figures in the region of 10%, often based on advice given by lenders at that time.

There is still a chance that interest rates could go down as well as up, so my advice to a farmer who can afford to pay an extra 2% interest rate is to use it to reduce the principal by 2%, not as a hedge against an increase in future rates. In my opinion the best hedge against an increase in future rates is to make a decision now to fix the amount in total that is paid in respect of a loan, rather than fixing the rate of interest.

Table 15.1 shows the number of years required to pay off a loan if the total payment per annum is respectively 10% and 12% and the actual interest rates vary from 7% to 9%. Fixing the annual payment at 12% pays off the loan in 16 years if the actual interest proves to be 9%, 14 years if 8% and in 13 years if the actual interest rate is 7%. If

Table 15.1 Number of years to pay off loan

Total payment per annum		*10%*	*12%*
Actual interest rate	7%	17 years	13 years
	8%	20 years	14 years
	9%	25 years	16 years

the annual payment is reduced to 10%, the time taken to pay off the loan is increased to 17, 20 and 25 years respectively.

Mention needs to be made here of 'real' interest rates, in other words, the amount by which the actual interest rates exceed the rate of inflation. At the present time, inflation is in the order of 2.5% and the rates charged to farmers tend to be in the region of 7.5%, giving an effective real rate of 5%. Interest rates have been at real rates ever since the Conservative Government was re-elected in 1979. Prior to that, during the 1970s, interest rates in effect were negative as the rate of inflation was more than the rate of interest.

It should be borne in mind that if interest rates should rise very substantially in the future, this would probably be due to a period of increased inflation and the real rate of interest would probably not increase. In other words, if one takes a decision to borrow at variable rates, one is not likely to find oneself at some future date worse than the average, but if one takes a decision to fix interest rates one could find, as was the case last year with the farms referred to earlier, that real rates are well above average. In these cases, action had to be taken to change from a fixed to a variable rate and this involved the payment of a penalty to the lender.

When borrowing money from a lender it is important to remember that the lender is in the business of 'selling' money. He wishes, obviously, to have a satisfied customer, but when the chips are down, the bank's or AMC's interest comes first.

Returning to the question of fixing the amount paid per annum, it is most important not to agree to pay off a loan more rapidly than is feasible. The strategy should be to aim to obtain a loan for as long a period as possible, and obtain agreement to be able to pay off this loan more quickly if circumstances allow.

Even if circumstances do allow the loan to be paid off more quickly than originally agreed, this may not be the best practice. In many cases it would be more advantageous to invest the surplus funds in a reserve account and pay off the long term debt only when the need for this reserve account becomes clearly unnecessary. So to state a maxim, 'Borrow for as long as possible but aim to be in a position to be able to pay the debt off earlier than agreed'.

PRINCIPAL REPAYMENTS ARE MADE AFTER TAX

The need to repay the principal (ie the amount of money borrowed, not the interest) out of taxed income is the major reason why loans cannot be paid off as quickly as anticipated. This simple fact is often forgotten by farmers, bankers and advisers alike. It is all very well to produce a table such as Table 15.1 which shows the period in which a loan can be paid off, but it has to be remembered that these tables do not take into account tax payments.

If, for example, a decision is taken to pay off a loan of £60,000 on a straight line basis, ie £6,000 per year, it needs to be remembered that in the case of a 40% tax payer, a payment of £4,000 tax would also be required. In other words, a surplus profit of £10,000 is required to pay off the debt of £6,000. For a tax payer at 25%, the profit required to pay off the debt of £6,000, is £8,000.

The second major reason why loans cannot be paid off as quickly as anticipated, is the capital intensive nature of dairy farming. The need for reinvestment in items such as machinery, equipment and buildings is nearly always more than the historical rate of depreciation shown in the accounts, and there is always the need to buy more milk quota. Hence the advice to place surplus funds into a reserve account before making a commitment to pay off a long-term loan.

This advice is given on the assumption that the farmer concerned is prudent and is not likely to reinvest surplus funds unwisely, eg in fast cars, big tractors and prestigious buildings. If this is the case, it could well be more prudent to repay the loan quickly so that these unnecessary investments are not made, or at least not made until such time as the loan has been paid off or considerably reduced.

NEGOTIATIONS WITH LENDERS

Prior to the meeting with a lender, it is important to get one's facts and figures together and have a well thought out strategy. Lenders like to see up-to-date audited accounts. They also like to see up-to-date management accounts but, for some reason, place much more emphasis on the audited accounts, even though these usually do not show as much detail and as true a reflection of the progress of the farming business as that shown in management accounts.

They also like to see cash flow projection and are most impressed when they also have actual cash flow results compared to budget. Surprisingly, however, none of the major bankers appear to have mastered the problem of producing suitable cash flow forms of their own on which the actual results can be compared to budget. The information required on their schedules is often too detailed and is based on comparison of month-to-month figures and does not allow true progress of the business to be measured, ie by comparing the cumulative cash flow to budget, as discussed in Chapter 11.

Lenders have rules of thumb in relation to their lending which are often based on a rent equivalent as a percentage of output. They add together interest charges, hire purchase payments, rent charges and loan repayment commitments to arrive at what they term a 'rent equivalent' and express this as a percentage of the output or gross margin: 25% would be considered to be fairly high, 10 to 15% would be considered more normal and easy to justify.

This percentage figure is a very useful rule-of-thumb measure; the higher it is the more closely the lender has to look at the proposal and if it is higher than normal he will require some specific reasons why he should provide the necessary funds.

Security is very important to lenders, and by security they mean land and property and insurance policies. Cows and milk quota make the profits but these assets are not regarded as true security. Milk quotas could disappear and the cows could die. Tenant farmers therefore have much more difficulty compared to owner-occupiers in borrowing capital. Rule-of-thumb figures are to lend up to 50% against the value of property subject to the finance charges being reasonable in relation to the gross margin.

Quota leasing charges are now an important part of what are, in effect, financial outgoings on a dairy farm, and these need to be added to the items above to arrive at a true rent equivalent. Bearing in mind the increasing significance of milk quota, it is more appropriate these days to work out what the total finance charges are and express them as a *quota leasing equivalent*. In today's climate this gives a more accurate rule-of-thumb measure of the viability of a dairy farming business than a rent equivalent per acre. Quota leasing charges are in excess of 6p per litre but very few dairy farmers can afford to pay a rent equivalent of 6p per litre and still leave a margin for reinvestment.

If the reader would like to study this in more detail it is suggested he or she examines the data shown in Table 10.1. This farm produces 700,000 litres. The rent and finance charges in 1992 were £20,000 or 2.9p per litre and there was no margin that year for reinvestment.

In the following year the rent and finance charges were £18,000 (2.56p per litre) and in this year there was a margin for reinvestment after deducting private drawings of £6,780, ie just under 1p per litre.

The results in 1994 are expected to be much better than in the previous year. The rent and finance charges are expected to fall to £14,000 (2p per litre) and it is also estimated there will be a profit for reinvestment, after deducting private drawings of £19,800 (2.8p per litre). In other words, even in this good year the farm would not be capable of servicing a finance charge equivalent to 6p per litre. The amount that the farmer can afford, however, is now approaching 6p per litre, and it would appear he could afford to pay, say, 5p per litre, ie a total rent equivalent of £35,000.

The good results being achieved at the present time pose a dilemma for the farmer, the banker and his consultant alike. In other words, are rent equivalents in the region of 5p per litre sustainable during the next few years? If they are, now is the time to expand and negotiate increased funding from lenders, in other words, increase cow numbers, purchase quota, improve buildings and set the business up now for the 21st century.

There are approximately 29,000 dairy farmers at the present time and this is expected to reduce to not more than 20,000 within the next 10 to 15 years. It is quite likely that those who survive will be those who make the necessary reinvestment and take the risk.

During the past two to three years there has been a hardening of attitude amongst bankers. Most banks made substantial losses through bad lending in the late 1980s/early 1990s and during the past two to three years have been aiming to improve their balance sheet by a combination of charging higher levels of interest rate over base and such things as management fees. As a result, too much of the time at the annual meetings with the bankers tends to be devoted to negotiating the management fees and the percentage over base, rather than the actual funding required by the business. Hopefully, this will prove to be a temporary problem and, in due course, farmers will be able to spend more time discussing the funding, always providing that there is a bank manager with whom this can be discussed.

The local bank manager is becoming an 'endangered species'. One of the executives of the major banks has been quoted as stating that computers are more efficient than managers! Time will tell whether this is a correct judgement.

FUNDING THE LAST TEN COWS

The most frequent borrowing mistake on many farms is over capitalisation in long-term assets at the expense of short-term requirements. It is not unusual to find a dairy farmer who has invested large sums of borrowed capital in new buildings and equipment and then finds he is unable to find the funds to purchase the cows required to use the facilities. Lack of cows means lack of cash flow to service the borrowings, and the final result, in certain cases, can be the need to sell the farm.

It is a truism that most of the gross margin made from the last ten cows is reflected in the 'bottom line'. In other words, if the average gross margin is £900 per cow, the last ten cows will produce an extra £9,000 gross margin and at least two-thirds of this, ie £6,000, is likely to be reflected in an increased profit margin. If milk quota has to be leased, the amount reflected in the bottom line will be less, but will still tend to be fairly substantial.

Lenders tend to draw a line over which they will not lend more capital to a business, including the last ten cows. The problem is that the last ten cows will cost approximately £12,000. They increase the profit by at least £6,000, but it could take up to two years to pay for the cows. If the bank refuses to provide the funds to purchase the cows, and a short-term loan is taken up to pay for these cows over a period of, say, ten months, this is not feasible and a cash flow crisis is the result. The message, therefore, when organising the capital side of the business is to make sure that adequate funds are available for the last ten cows.

If they are not, do not borrow £12,000 and aim to fund it over too short a period. Sell, say, 20,000 litres of milk quota to pay for the cows, lease in the quota required and make sure that the profit made by these cows is available within two years to repurchase the quota.

The question of drawing a line referred to above can be taken to extremes by bankers in a way that seems difficult to justify, except on the principle that a line has to be drawn somewhere. The author has known more than one occasion when banks have refused to honour a cheque in payment for the leasing in of milk quota, knowing full well that if they did honour the cheque the Milk Marketing Board would release to the farmer monies owed to him for milk, well exceeding the cheque required to lease in the quota. This poses a real crisis for the farmer concerned. In order to survive, leasing in this quota is a must, so he has to turn to other sources of funding in times of crisis, that is the merchants and other people with which he does business, including his milk quota agent.

These cases really illustrate, I believe, the point that was made earlier in this chapter, that when the chips are down, bank managers work for their bank, not for the customer.

CHAPTER 16

Case Histories 1985/94

INTRODUCTION

These case histories are based on actual farms with which the author is closely involved and have been selected to illustrate the main themes mentioned in the preface to this book, which are:

1. 'Make the quota fit the farm, not the farm fit the quota.'
2. 'High fixed costs, not high feed costs, are today's main problem.'
3. 'Dairy farming is opportunist, and a gamble.'
4. A belief that a year would come when there would be no cut in milk quota. This would lead to a shortfall in cow numbers and a substantial increase in their value.
5. A belief that milk quota values would tend to be a hedge against inflation, at least until the year 2000.

The reader is reminded that cow values fell very considerably when quotas were first introduced and a newly-calven cow could be purchased for no more than its cull price.

The author well remembers a £5 bet that was made with a client in 1988 that cow values would reach £700 in 1989: the client was quite happy to pay his bet in 1989 when cow values reached £700, as a decision had been made in 1988 to expand the herd before the price went up.

DAIRY FARM PARTNERSHIP

This case history concerns a business that was set up in April 1985. The opportunity for the partner came because the business at that time was mid-way through a development plan to increase cow numbers and this came to an abrupt end with the introduction of quotas.

Milk production / quota

The quota on this farm today, if no quota had been purchased or leased, would be approximately 590,000 litres. The actual production in the year ended 31 March 1994 was approximately 1.14 million.

During the first three to four years of the partnership, milk quota was leased, not purchased, as at that time the interest on the capital value of milk quota was very similar to the leasing charge, in other words, in the region of 5p per litre.

Leasing prices increased significantly in 1989/90 to a figure in the region of 6.5p per litre and this was well in excess of the interest charges on the capital cost of quota, which were in the region of 33 to 35p.

At that time, the saving of 1.5 to 2p per litre in finance charges compared to the cost of leasing was very important in proportion to the profit that could be expected at the end of the year. The bank was therefore approached in 1990 to see if they would consider funding the purchase of milk quota. Eventually, a deal was struck on the understanding that some of the milk quota purchased would be sold off at a future date to prove to the bank that milk quota was, in effect, something that could be sold.

Part of the quota was sold, as agreed but with some misgivings, in the spring of 1992 at 29p per litre. Fortunately, the bank agreed to fund the re-purchase of a similar amount in the autumn of the same year and this was done at 27p per litre.

Some of the quota debt has now been paid off out of cash flow. It is comforting to know that the price of quota now is in the region of 47p per litre and the debt could be paid off by selling only a small part of the quota originally purchased. Whether or not this is done will depend in part upon the next meeting with the bank manager and how keen he still is to see this debt repaid.

Livestock numbers

The partnership started with 112 cows and 70 youngstock. There are now 200 cows in the herd. These are on the herd basis, the extra 88 having been transferred into the herd at cost of rearing.

Cow numbers in the quota year 1993/94 averaged 182, giving an average yield of 6,250 litres per cow. This is quite a modest yield but it is achieved from a simple self-feed silage system with no facilities outside the parlour to feed concentrates.

The strategic objective formulated four to five years ago was to aim to have a herd of 200 cows so that 20% could be sold off on the herd basis, leaving a herd of 160 cows capable of producing one million litres (6,250 litres per cow). When this strategy was formulated in 1988/89, it was hoped that the 40 cows would realise in the region of £700 per head, ie £28,000, enough to pay the purchase price of 80,000 litres at 35p.

Forty cows could probably be sold during the coming year for a figure in the region of £40–48,000, but there is some doubt as to whether this should be done as it would lead to a cutback in production and this could reduce the potential farm profit. There is also less pressure to reduce the bank borrowing when interest rates are in the region of 7 to 8%, compared to 15% as recently as two years ago.

Youngstock numbers at the present time are in the region of 70, very similar to those when the business was first started in 1984. Four years ago a decision was taken to use a beef bull on all the cows over a period of approximately 12 months. The income received from the beef calves has aided cash flow and the build-up in cow numbers to 200.

A decision has now to be made as to whether a herd reduction sale should take place this year as originally planned, or whether future plans should include an increase in the number of youngstock reared, the objective being to have a herd size of, say, 240 in 4 to 5 years' time so that 20+%, ie 48+, can be sold on the herd basis in 1998/99 or 1999/2000, ie after the surplus of heifers already in the pipeline on other farms has worked its way through the system.

Staffing

Two full-time men have been employed on the farm throughout this period and a major objective has been to contain the labour, plus machinery costs per litre.

This was only feasible in the early years by leasing in additional milk quota. Leasing charges at that time represented a very high proportion of the profit and based on the trading result it appeared difficult to justify.

It was, however, anticipated that one day there would be a capital appreciation in the value of the dairy herd and this would justify the strategy. Actual events have now proved this to be the case.

Improvements

As mentioned earlier, the farming system is based on self-feed silage and emphasis has been placed on keeping labour and machinery costs to a low level. Investment in field machinery has been kept to an absolute minimum and the silage has been made on contract.

The parlour was updated from a 10:10 to a 16:16 in 1986. Priority was given to updating the parlour at that time as it was essential in order to achieve the labour objective identified above, and at that time, ie immediately after the introduction of quotas, there was no difficulty in obtaining competitive quotes to do the work.

It is now nearly 10 years since the parlour was updated and the need for a further update is imminent.

The dirty water system was updated approximately two years ago to bring it in line with the National River Authority's requirement. This expenditure was on a stop/go basis for a period of three years before a satisfactory solution was found. As with all farmers, the capital expenditure incurred for this item was made with some reluctance, as in effect it was simply an investment to stay in business and did very little to improve the intrinsic efficiency of the business. Experience, however, has shown that the dirty water is effective as a fertiliser; good grass crops have been achieved following its application.

Improvements are being considered which will allow feeds other than grass silage and brewer's grains (clamped under the silage) to be introduced into the diet, the objective being to increase yield per cow.

In 1993/94 1.1375 million litres were produced with 182 cows at 6,250 litres per cow. The objective by, say, the year 2000 should be to produce, say, 1.16 million with 160 cows at 7,250 litres per cow.

It was said at the start that the management objective for this business was to contain labour and machinery costs; the extent to which this has been achieved can be judged from the information shown in Table 16.1.

During the nine year period covered by the results shown in this table, the average price received for the milk has increased from 14.61p to 21.10p per litre, ie an increase of 6.49p. Most of this has been reflected in an improved margin over feed cost per litre, which has increased during the same period by 6.35p per litre. Much more important, however, is the control that has been exercised over labour and machinery costs. The margin after deducting labour and machinery costs, as well as feed costs, is up by 5.62p per litre.

Milk production during this nine year period has increased from 736,652 litres to 1,135,840 litres or by 54%. During this same period

Table 16.1 Financial results—dairy farm partnership

	Margin over purchased feed*		Labour and machinery costs**	Margin over feed, labour and machinery	Number litres produced	Per litre		
	Budget	Actual				Milk price	Margin over feed	Margin over feed, labour and machinery
	£	£	£	£	No.	p	p	p
1985/86	69,600	77,115	36,515	40,600	736,652	14.61	10.47	5.51
1986/87	88,000	95,962	35,214	60,748	875,202	15.13	10.96	6.94
1987/88	90,000	93,745	29,927	63,818	824,284	15.78	11.37	7.74
1988/89	108,000	112,759	46,466	66,293	870,914	16.84	12.95	7.61
1989/90	124,000	127,270	46,816	80,454	928,445	18.32	13.70	8.67
1990/91	136,000	129,858	53,179	76,679	939,474	18.74	13.82	8.16
1991/92	145,000	144,648	60,007	84,641	998,749	18.86	14.48	8.47
1992/93	156,000	163,007	60,028	102,979	1,098,781	20.02	14.83	9.37
1993/94	174,000	191,080	64,605	126,475	1,135,840	21.10	16.82	11.13

* Including purchased bulk feed as well as concentrates.
** Including depreciation.

the margin over purchased feed, labour and machinery costs has increased from £40,600 to £126,475, or by £85,875. The margin over purchased feed is up by £113,965, the increase in labour and machinery costs is only £28,090.

The existing system has worked well but the business is now at a crossroads. Changes to the farming system and further improvements in the total milk yield can only be achieved with significant capital expenditure on improvements.

Possible strategies that could be adopted have already been discussed briefly. These will have to be looked at in much more detail. A decision will be made after detailed budgets have been prepared and these have been discussed with the bank manager.

BUYING QUOTA

This case study has been included to illustrate the opportunist nature of buying milk quota.

Prior to 1991/92 the farming system on this farm was based on the production of 1.1 million litres of milk per annum and plans were prepared to purchase 50,000 litres of milk quota in 1991/92.

For various reasons, the purchase of this quota did not go ahead as planned. A meeting was held with the farmer's banker in November 1992 to discuss the budgets for 1992/93, including the purchase of 20 additional cows and 50,000 litres of milk quota. As usual, the management accounts for the previous year were discussed in detail together with the budgets for 1992/93. The plans put forward were approved—end of meeting? There was time for a 'bit of chat' before the bank manager moved on to his next appointment.

The bank concerned were advertising that they were prepared to finance the purchase of milk quota and it seemed a good idea to discuss this with the bank manager to see 'in theory' how much finance they would be prepared to put up to purchase milk quota on this particular farm, the idea being that if it was purchased it could either be produced or leased out.

The price of milk quota at that time was particularly favourable, ie only 27p per litre, and the bank manager indicated that he should be able to put a case up to head office to fund the purchase of, say, 370,000 litres at a total cost of £100,000. End of discussion.

Next morning, the farmer telephoned to say it seemed more than just a good idea. Please contact the bank manager, put the case to head office and see if £100,000 funds can be obtained. The answer was, 'yes'.

The problem then was to procure this quota without putting up the price. The milk quota market is very volatile and a sudden demand from one producer to purchase nearly 400,000 litres could lead to a significant increase in the price. Vendors of milk quota do not have a very good track record for sticking to verbal deals as it is not unknown for vendors to renege on verbal deals if the price goes up. Verbal deals were agreed to purchase milk quota to the value of £100,000. One of the deals fell through. The amount eventually purchased was just over 310,000 litres at a total cost of £84,000 (27p per litre).

A loan was then arranged with the bank for £84,000 and it was agreed that £1,250 per month would be transferred from the current account to the loan account to cover interest and principal repayments, the objective being to pay off the loan in 7 years. This loan payment of £15,000 per year is equivalent to 300,000 litres at 5p per litre, the price at which it was envisaged quota could be leased out if the milk was not produced.

It is now March 1994, milk production is above quota, and looking back the question asked is 'Why didn't we buy more?'. The objective in business, however, is to look forward. The opportunist purchase of this quota and the decision taken to expand cow numbers to produce the milk provides a springboard for further expansion. Tentative plans are now being prepared to expand production further, the objective being in due course to produce 1.8 million litres.

This example shows bankers in a good light but they are not always so co-operative. The farmer in this case was well established, he could make profits without the purchase of the additional milk quota, he had a good asset base and the lending represented virtually no risk to the bank. Both the farmer and the banker were almost certain to make profits out of the deal.

A few months earlier, in the summer of 1992, a very similar proposition was put to a different banker. In this case the farm was producing 1–1.1 million litres per year but had a quota of only 650,000 litres, and also had a fairly high level of borrowing. In other words, it was 'the last ten cows syndrome'—the business needed to borrow more money to ensure future success.

A proposition was put to the bankers that £100,000 should be made available to fund the purchase of 350–360,000 litres to ensure the continued viability of the business. In this case, the answer was no, so the future viability of this business is still in question.

Provision has to be made in 1994/95 to lease in 350,000 litres of milk quota. At 6p per litre, this is £21,000, and if the cost should rise to, say, 8p, the cost of leasing in this quota would be £28,000.

The interest charge on a £100,000 loan would be no more than £8,000. The business would also have an asset worth £70,000 more than its purchase price. It would seem fair to say that these two case histories illustrate the old saying about bankers, *'They are very happy to lend you an umbrella when the sun is shining, but not when it is raining!'*

Perhaps more important, it also illustrates that in dairy farming, as in any business, one has to back one's own judgement. Tremendous profits are there to be made if you are right, and everybody else is wrong. It takes courage, however, to march out of step with your fellow-farmers. This has certainly proved to be the case in the author's experience of milk quota: the right time to either buy or lease in milk quota has always been at a time when the majority of one's fellow-producers did not think it was necessary.

It is often said that the difference between a good farmer and a bad farmer is a week! The difference between getting it right and not getting it right with a milk quota transaction is in the region of four to six weeks. *If you leave your decision to lease in or buy quota until you are sure it is right, it will be too late!*

REDUCING BANK BORROWING

The following case history has been included as it is important to learn from mistakes as well as to build on successes.

This farm had a creeping overdraft (upwards!). This was giving cause for concern to both the farmer and his bank manager. Decisions were therefore taken in 1992 that the business should be reorganised and that a major objective should be to reduce the bank overdraft. A strategy was eventually agreed which placed the emphasis on increasing the yield per cow and reducing the labour and machinery costs per litre.

A decision was taken for cash flow reasons only (in other words, to reduce the overdraft) to sell youngstock. This decision was made with some reluctance as it was thought at the time that the youngstock could increase significantly in value during the coming months, as, in fact, they did. Some of the proceeds of the sale, however, were to be invested in dairy cows in order to increase the MoC.

Budgets were prepared for the quota year 1993/94 on the assumption that additional cows would be purchased in the quarter January to March. Again, for cash flow reasons, these cows were not purchased. It was hoped that the budget objective would be achieved without buying these cows, but in the event this proved not to be the case.

A substantial improvement was achieved in both the milk yield and the margin per cow, but the total MoC fell below budget. There was a tendency to have to chase quota with consequent increase in feed costs. A decision was eventually taken in November to purchase the cows that should have been purchased 9 months earlier.

This lesson, however, has now been learned. Plans for 1994/95 are being prepared on the assumption that cows will definitely be purchased before the end of March 1994 if this is necessary to achieve the budget MoC for 1994/95.

EXPANDING PRODUCTION/ CONTAINING FIXED COSTS

This case history starts in December 1987. The farmer concerned owned approximately 100 acres, including 30 acres purchased just before the introduction of milk quotas, and 'rented' 15 to 20 acres. At that time, ie 1987, the milk quota owned was 462,000 litres and milk production was in line with quota, in other words, the farm policy was being made to fit the quota available.

Youngstock were being reared and the management objective was to produce as much silage as possible, as was fashionable at that time, the idea being that this would reduce the cost of purchased feed and improve margins. In other words, the farmer was following the perceived wisdom of most advisers and commentators at that time in relation to the problems imposed by milk quotas.

The farmer and his wife 'did not feel the business was going anywhere'. They had a young family and hoped that in due course the business would provide a living for both themselves and the family. When they first started farming they kept 80 cows on 60 acres and bought in feed. Professional advisers tended to criticise the business for its high feed costs and this was the philosophy of the current adviser.

The accounts were examined for the six years ending March 1987, ie 1982 to 1987. They revealed that the fixed costs on this farm were much lower than normal and because of this the farm had managed to make reasonable profits in all six years, despite the cutbacks made after the introduction of quotas.

They also revealed that the highest profit was made in the year ended March 1984. The total milk sales in that year, ie 1983/84, were £5,000 more than they were in 1986/87. Feed costs were lower in 1986/87 than they were in 1983/84 but the farm was still making less money than it did in 1983/84, which is on record as being one of the worst years for dairy farming, owing to a combination of wet weather and high feed costs relative to milk prices.

It was agreed the business was not going anywhere and that a radical change in strategy was required. The average number of cows in the herd for the year ended 31 March 1988 was 92, and there were 50 youngstock on hand. It was agreed that the number of dairy cows should be increased to 140 as quickly as possible. In the short term, milk quota would be leased but it was hoped that funds could be found to purchase milk quota in due course. Control of fixed cost was to remain an important priority and, if possible, the plan was to be implemented without any significant increase in borrowed capital.

It was recognised that keeping 140 cows on just over 115 acres would mean that bulk feed would have to be purchased, but it was agreed this was a key part of the strategy, as it was considered bulk feed could be purchased for not more than its cost of production.

The extent to which this strategy has been successful can be judged from the information shown in Table 16.2. In the two years ended 31 March 1988, the margin over purchased feed averaged in the region of £56,000. After deducting fixed costs there was a margin over feed and fixed costs averaging £22,454.

(Note: To arrive at the profit, the forage costs have to be deducted from this figure and adjustments have also to be made for the change in livestock valuations and the contribution made by the

Table 16.2 Case history farm results 1980/87 (farm A)

Years	*Cow nos.*	*Litres produced*	*Margin over purchased feed*	*Fixed costs**	*Margin over feed and fixed costs*	*Quota leasing charges*
1986–87	89	496,353	56,791	32,463	24,328	Nil
1987–88	92	490,832	55,247	34,666	20,581	Nil
1988–89	113	601,690	73,787	43,082	30,705	3,875
1989–90	135	711,040	92,463	55,476	36,987	13,184
1990–91	141	780,621	102,898	54,119	48,779	11,232
1991–92	136	792,992	112,060	58,551	53,509	7,475
1992–93	139	824,972	121,461	56,451	65,010	1,335
1993–94	145	890,891	134,799	66,600	68,199	1,600

* These include labour, power and machinery, rents and rates, property repairs, miscellaneous overheads, depreciation and finance charges.

youngstock enterprise. This information has purposely not been given as it is not wished to reveal the actual profit made by the business.)

Cow numbers reached the desired level of 140 by the end of 1990 and averaged 141 in the year ending March 1991. The margin over feed cost by the end of March 1991 was approximately £47,000 more than that achieved in 1987/88 and by the end of March 1992 it reached £112,000, double the 1987/88 figure. The margin over feed cost in 1993/94 is expected to total £135,000 from 890,000 litres. The objective in 1994/95 is to achieve a margin over feed cost in the region of £156,000, that is £100,000 more than that achieved in 1987/88.

Fixed costs excluding quota leasing charges have risen as a result of the increase in cow numbers and inflation but were remarkably constant for the four years ending March 1993. During this four year period they were contained at approximately £56,000 and as a result, the margin over feed and fixed costs increased from approximately £37,000 in 1989/90 to £65,000 in 1992/93.

During the three years 1989 to 1991, priority was given to increasing cow numbers and no milk quota was purchased, consequently a significant part of the additional margin achieved in these three years was absorbed by high quota leasing charges.

Advantage was taken of the low quota prices in 1990/91 and 1991/92, to purchase 325,000 litres, bringing the total quota owned to just under 790,000 litres. As a result, quota leasing charges in 1992/93 and 1993/94 averaged only £1,467 compared to a high of £13,184 in 1989/90.

The margins being made in 1993/94 are now being used to fund other capital items, such as improvements to buildings and fixed equipment. There has been a need for these investments for several years but the business philosophy is to spend on these items only when the cash becomes available. This philosophy has certainly paid off.

An outline strategic plan is now being prepared for the next three to four years. This is likely to be based on a herd of 160 cows, hopefully producing in excess of one million litres. This will lead to an even tighter stocking rate, but it is not planned to increase the area farmed.

HIGH MoCs DO NOT GUARANTEE HIGH PROFITS

This case history is exactly the opposite to the previous one. In this case the management philosophy has been to aim to produce high MoCs per litre and high MoCs per cow. The extent to which this has been successful can be judged by examining the data shown in Table 16.3, Farm B. The margins per litre and per cow achieved by farm B are much higher than those achieved by farm A (previous case history), but unfortunately farm B is not profitable, as too much has been spent over the years on forage costs and fixed costs to produce these excellent margins. This business is going backwards. The total MoC in 1993/94 is £32,197 less than in 1991/92, even though the MoC per cow has gone up from £1,008 to £1,248, and the MoC per litre from 16.53 to 19.11p.

The problems on this farm illustrate the most important concept that one needs to grasp in relation to dairy farm business management. The objective is to produce the maximum profit from the

Table 16.3 High MoCs do not guarantee high profits

	PROFITABLE FARM (A)					*LESS PROFITABLE FARM (B)*				
Year	*Total margin over feed (£)*	*Margin per litre (pence)*	*Yield per cow (litres)*	*Margin per cow (£)*	*Cow numbers*	*Total margin over feed (£)*	*Margin per litre (pence)*	*Yield per cow (litres)*	*Margin per cow (£)*	*Cow numbers*
1988–89	73,787	12.26	5,324	653	113	79,316	13.99	5,609	785	101
1989–90	92,463	13.00	5,266	685	135	114,089	15.94	6,016	959	119
1990–91	102,898	13.18	5,536	730	141	114,244	16.99	5,355	910	125
1991–92	112,060	14.13	5,830	824	136	127,036	16.53	6,099	1,008	126
1992–93	121,461	14.72	5,935	874	139	107,289	17.77	6,306	1,119	96
1993–94	134,799	15.13	6,144	930	145	94,893	19.11	6,530	1,248	76

whole farm, after deducting fixed costs, not the highest margin per litre, margin per hectare or margin per cow.

If you study the information for the previous case history, farm A (Tables 16.2 and 16.3) you will find that the margin per litre is not particularly good and neither is the yield per cow. Fixed costs, however, have been very well controlled. The net result is a substantial profit margin and a steady growth in the size of the business.

The results for farm B are also shown in Table 16.3. If one only examined the MoC per litre and per cow the conclusion would be that this farmer is doing an excellent job. The yield per cow, the margin per litre, and the margin per cow are all significantly better than those achieved by farm A. Farm B, however, is not successful as the cost of producing this margin, ie the fixed costs plus forage costs, are too high.

This need to look at fixed costs as well as gross margins has made the writing of Chapter 5, concerned with margins, exceedingly difficult as these had to be discussed without any detailed reference to fixed costs.

The conclusion to be drawn is that if you want to make profits from dairy cows you have to ensure that it does not cost you too much in the way of fixed costs and forage costs to produce the MoCs per cow and per litre. In other words, *it is not the margin per litre that matters; what matters is what it cost you to produce the margin and the difference between the two!*

CHAPTER 17

Retiring from and Getting Started in Dairy Farming

GETTING A START

Getting a start in dairy farming has always been difficult, and the introduction of milk quotas has certainly not lessened these difficulties. It has, however, provided a means whereby some farmers who were having difficulty in finding enough cash to retire have been able to do so by either selling or leasing out their milk quota.

When one is young and enthusiastic one is very keen to obtain a tenancy or any rung on the farming ladder. It is, however, pertinent to note that after, say, 30 years in farming some of today's tenant dairy farmers do not have enough capital on which to retire. In some cases father has enough capital on which to retire but does not have enough to provide sufficient capital both for himself to retire and for his son to start up in business on his own account.

How much capital you require to start in farming, or to farm, is a topic that is often discussed and is one on which many farmers, teachers and advisers would have a view. How many, however, would be in a position to answer the question 'how much capital do you need to retire?'

Professional advisers and teachers are usually salaried and think in terms of a pension based on their number of years' service and a percentage of their retiring salary. It is not often discussed in terms of a capital sum.

A useful rule-of-thumb is that one needs £100,000 to provide a pension of £10,000 per annum, based on retirement at the normal age of 65 years. The pension sum can vary by plus or minus 25%, depending upon whether widow's benefits are included and the time at which the annuity pension is actually taken.

That's enough for the time being about retiring, let's go back to the question of getting started in farming.

LEARNING THE JOB

To succeed in dairy farming, you must be good at milking cows and caring for cows. To acquire the necessary skills there is no substitute for practical experience. This practical ability to do the job on the farm is essential if you want to work your way up the farming ladder.

A lot can be learned from watching how other people approach and do the job, but a young person needs to avoid simply getting the same experience ten times over. This can only be achieved by working on more than one farm, and this need to work on another farm is particularly vital in the case of a farmer's son.

This, however, is not always possible. Experience has shown that farmers' sons and daughters gain a tremendous amount from attending day-release or other short-term courses, where they rub shoulders with their peers and learn new ideas. Experience has also shown that this may be a better way of acquiring experience if one definitely wants to farm, rather than attending a full-time course. There is a danger that on a full-time course one would lose the 'work ethic' that is so important in farming.

This ability to do the job and care for the animals has to be coupled with the ability to run the farm as a business. It is fair to say that this will not always be a skill that comes readily to a person who is much more interested in the day-to-day management of the farm, which explains why the role of the farm business consultant has become much more common on dairy farms during the past 20 to 30 years. It is difficult to imagine dairy farming at the present time without an organisation such as Genus Management, but this organisation was very much in its infancy and on trial when the author was lecturing to students some 30 years ago.

The difficulty for the young entrant is that the fees charged to him or her tend to be the same as for a larger farm, but it is most important to have a viable business plan from the start. If not, a young person could find that he or she has worked very hard and at the end of the day has no reward to show for this hard work as the original business plan was not viable.

CAPITAL AND SIZE OF FARM REQUIRED TO START FARMING

Immediately, it needs to be stressed that the amount of quota available is probably more significant than the size of the farm. The amount of capital required will also depend upon how little the person starting up in farming can manage to live on. Success in building up a dairy farming enterprise, like any other business, depends on one's ability to save and reinvest in the business, no matter how small the profit may be.

This comes down to such basics as how much money to spend on the kitchen! £5,000 spent on a kitchen, instead of cows, in the first year of a dairy farmer's business plan could be the make or break decision determining the viability of the business in the long term.

A useful rule-of-thumb for determining the size of business required is that the gross margin from the business should be three times that of the agricultural wages of a cowman, as the fixed costs, other than labour, tend to be approximately two-thirds of the gross margin. Given this rule-of-thumb, we can calculate the size of farm and the amount of milk quota and capital required, as shown in Table 17.1.

This budget is based on a farm of 25 hectares with a milk quota of 250,000 litres produced by 40 cows at 6,250 litres per cow, and shows a potential profit of £16,500 before tax and drawings, which is 42.8% of the budget gross margin. This is more than the rule-of-thumb of one-third, partly because dairy cow profitability is relatively high at the present time (1994), and the rent and quota leasing charge assumed in the budget, at £7,500 (equivalent to 3p per litre) may prove to be optimistic in practice.

The assumed rent of £300 per hectare (£120 per acre) may not be enough in a competitive environment if the farm has 250,000 litres of milk quota. It is quite possible that a rent of this magnitude would have to be paid even if the quota available was only 200,000 litres, leaving an additional 50,000 litres of milk quota to be leased at a cost of say £3,000 (6p per litre), reducing the budget profit margin to £13,500 (35% of the gross margin).

It is stressed that these are budget figures. The actual results achieved would depend very much on the farm and the skills of the particular prospective farmer.

The table also gives an indication of the capital that would be required to set up this farm. The budget total is £84,500, including an allowance for the ingoing and tenant's fixtures paid to the outgoing tenant, of £11,000.

Table 17.1 Break-even budget for a new entrant (at 1994 prices)

		£	£
Gross margin:			
Margin over concentrates 250,000 litres @ 16p (40 cows @ £1,000)			40,000
Calf sales (35 born, 10 retained)	25 @ £160		4,000
Cull cows	10 @ £450		4,500
			48,500
LESS youngstock feed 25 @ £80		2,000	
Vet. and med. and sundries—youngstock		1,000	
—cows		3,000	
Forage costs—25 hectares @ £160		4,000	
Bulk feed (adequate area rented)		Nil	10,000
			38,500
Fixed costs:		*Per hectare*	
Power and machinery		400	10,000
Repairs and maintenance		60	1,500
Sundry overheads		120	3,000
Rent and quota leasing charges 25 hectares @ £300			7,500
			22,000
PROFIT MARGIN before tax and drawings			16,500
LESS drawings and tax			12,000
MARGIN for reinvestment			4,500
Capital requirement:			
Dairy cows	40 @ £1,000		40,000
Youngstock	25 @ £500		12,500
Machinery and equipment 25 hectares @ £600			15,000
Tenant's fixtures			5,000
Ingoing			6,000
2 months' working capital			6,000
			84,500

The tenant's fixtures/ingoing will depend very much upon the nature of the farm being taken over. This could be either a high or a very low figure; £12,500 is included for the capital that would be required for the youngstock and £15,000 for the machinery.

At the start, the investment in youngstock would probably be nil, but the cows would have to be purchased as down-calved heifers or cows and would probably cost in the region of £1,200 per cow, ie

£48,000 compared to the budget figure of £40,000 shown in the table.

Looking at these figures, a prospective new entrant is not going to start farming if he is looking for a 25% return on his capital, as well as a payment for the work that he undertakes. The budget profit, *before* taking into account the value of his or her manual labour, is only in the region of 20% of the working capital requirement.

This re-emphasises the point made at the start of the book. If one wishes to farm, one does so because one wants to farm, not because one wants to make a good return on capital. A good return on capital, however, can be expected once the business is established and there is also the chance of the capital gain on the value of the cows, but a young person starting farming in 1994 will have to wait some considerable time before they can expect an appreciation in their stock values.

In this sense, the time to have started farming, that is if one could have had the opportunity, was approximately three to four years ago, when cows were under-valued in real terms and there was a potential for values to increase. History has shown, however, that cow values have tended to keep up with inflation over the years. Even at today's rate of inflation of 3%, one can expect cow values to double in the next 25 years.

OBTAINING THE NECESSARY CAPITAL

It has often been said that there are three main sources of capital for farming: parsimony, patrimony and matrimony!

1. Savings

Although, at first sight, it may not seem feasible, young people still manage to start farming where the basis of their capital is savings, not inheritance. As mentioned already, the prerequisite for such a young person wishing to start farming is the willingness to take less out of the farm in terms of cash than the value of the work they put in.

Inflation reduces the cash value of savings, so it is important to know how to invest savings as well as being able to make them. Buying a house is probably one of the best ways of saving capital, but this is not usually a feasible proposition for most herdspersons or herds managers as they are required to live on the farm. They do, however, have a house which is normally provided free of rent.

The amount of capital required to start up farming is usually very similar to the amount that is required to buy a house and this is something they will require one day when they retire. Therefore, a young couple who wish to farm should save as if they were trying to buy a house.

The first step in such a strategy would be to aim to save the deposit, ie 10% of the house price, representing a sum of say £8–10,000. This, it is suggested, should be the first objective as it will test the young person's ability to save out of income.

2. Part-time farming

Having saved some cash, part-time farming offers the best means of starting up, by renting either land and/or buildings on a short-term grazing licence. Beware, however, the threat that part-time farming may be to your job: Employers become rather anxious when they see their herdsperson dashing off to attend to their own enterprise.

Some form of part-time venture is very desirable before anyone contemplates starting up a farming business on their own account. It is much better to start initially on a part-time basis and come to realise the nature of the *risks* inherent in any business operation. Losses are made as well as profits, and often these occur in the initial years.

It is important to realise that if you buy five cows for, say, £1,200, one of them could die within a year and another could prove to be not in calf, leaving the need for two replacements and no profit in the first year.

3. Ownership of stock on the employer's farm

This is an arrangement that can work well for both the employee and employer alike, providing that both are aware of the mutual benefits they could derive.

Ideally the herdsperson or herds manager should start off with permission to rear a few calves with a view to these entering the owner's dairy herd at some future date. This fits in with the concept of developing the ability to save out of income, and also has the tax advantage that when the heifers calve down they will enter the employee's herd at the cost of rearing.

The heifer calf should be reared along with those of the owner and payments for the direct costs of rearing these calves, ie feedingstuffs, veterinary fees, etc, should be made from the employee to the owner. Rearing the calves along with those of the owner also ensures that they do not receive any preferential treatment.

At maturity, the heifers should be allowed to enter the dairy herd of the owner and the employee should then receive a percentage of the margin over concentrates for the cows he has in the owner's dairy herd. A way in which this can be arranged to the satisfaction of both the employer and employee is described below:

(i) The employee is allowed to provide up to 10% of the cows in the herd by introducing his own down-calving heifers.

(ii) The calf born to the employee's heifer is credited to his/her account and he/she is responsible for the veterinary costs and AI costs of getting a cow in calf.

The employee needs to receive the value of the calf to offset his potential herd losses and cover his herd depreciation costs.

(iii) The employee receives 12–15% of the margin over concentrate feed cost per cow per month, multiplied by the number of cows he has in the herd, to give him a reward for the capital he has invested in the cow.

(iv) The margin over concentrates of a typical herd is in the region of £1,000, so the employee would receive between £120 and £150 per annum, representing the return on his working capital, ie a down-calved heifer worth approximately £1,200.

(v) This payment to the employee is more than the interest charges that the employer will save by not providing the cow himself, but the employer benefits through the great degree of attention given to all the cows in the herd as a result of the employee's motivation.

The employer also has the satisfaction of knowing that his employee is less likely to leave, and knows that he/she is building up a capital sum that will be useful in the future.

4. Acquiring the farm and the quota

The lack of farms to let is the biggest stumbling block to the would-be dairy farmer. The decision of various County Councils to sell off their agricultural farms has also contributed to this problem, reducing the number of County Council tenancies that are available.

A few would-be farmers will get their first foot on the farming ladder by this traditional method of obtaining a County Council holding. Competition for these, however, is getting more and more fierce as the years progress and the would-be farmer needs to start to look at other opportunities such as share farming and contract farming. These offer less security than the traditional tenancy, but this is the very reason why they are becoming more popular. The would-be farmer, therefore, has to be prepared to take the risk and have a good look at the share farming/contract farming opportunities.

SUCCESSION

Most young people starting farming will be doing so by means of inheritance. At first sight, there may appear to be no problem as far as the farmer's son or daughter is concerned, but problems are almost bound to occur unless the succession is planned with care.

In the case of tenant farmer's sons and daughters care has to be taken to see that the tenanted farm continues to be the main means of livelihood for father, as well as son or daughter. If not, succession to tenancy could be questioned. For example, the purchase of a substantial area of land could put the succession to tenancy at risk, especially if separate businesses are established.

Sons and daughters, therefore, on farms where father has a 'three-generation' tenancy need to discuss with their father, if necessary, the importance of ensuring that steps are *not* taken that will jeopardise their chance of succeeding to the tenancy. They themselves also need to take steps to try to ensure that they will be suitable tenants by creating a favourable impression with the landlord. If this is achieved it may be possible to get the landlord's approval in principle to their succession well before the time it is expected to take place. This can then make the planning of the development of the future business much more simple. For example, if there are two sons, one can take the tenancy and the other can then seek to establish a business elsewhere on his own account, eg by purchasing a small farm. This is not theoretical: in a previous chapter reference was made to a tenant farmer's son who had purchased a farm. He did this in the knowledge that his brother would succeed to the tenancy of the existing holding.

There are also problems on the solely owner-occupied farm where there are several sons and daughters and some want to farm, and some do not. Parents are then confronted with the problem, do they gift most of their estate to one or two sons or daughters who wish to farm, or do they wish to divide their estate equally between their children. If they decide on the latter, it could well mean that the whole of the business has to be sold.

Discussions need to take place at an early stage in circumstances such as these. Often it will not be possible to give equal shares to all the sons and daughters and parents may need to place more emphasis on arranging their affairs in such a way as to give each child an equal opportunity of an adequate income during their lifetime. For example, a daughter may wish to establish a non-farming business, and in this case it may be appropriate to gift to her a significant amount of capital well before the son, who wishes to farm, inherits the farming business. Parents who have long-established businesses tend to see themselves as custodians of the business during their lifetime and their main objective is to pass it on to the next generation intact. To be fair, however, in most instances some provisions have to be made for sons and daughters who are not going to continue with the farming business. This is obviously a difficult problem and it is of utmost importance that all members of the family are fully aware and, if possible, informed as to why family funds are being allocated in the way they are.

A particular problem, these days, is that father is often in good health when his son or daughter is keen to take over the business and is not ready to step down and retire at the ripe old age of say 58. In this instance, means generally have to be found whereby the business can be expanded so there is a role for more than one member of the family.

Sometimes the solution is for the son or daughter to take over the home farm with father moving out to a separate holding.

SUCCESSFUL RETIREMENT

A theme of this book has been 'making it happen' and at some stage in one's life one has to begin to think in terms of retirement. It would be fair to say that one of the best measures of success is the ability of a successful dairy farmer to retire and leave the business he has built up in the capable hands of his successor(s).

To do this, he has to start to make plans well before retirement to overcome various obstacles, including those itemised below:

1. Inheritance tax

This is no longer quite the obstacle that it was, as since 10 March 1992 100% tax relief has been given on the inheritance of agricultural land, providing that it is owner-occupied. For tenanted land the relief is 50%.

Inheritance tax is payable at 40% on transfers in excess of £150,000 after the various reliefs have been deducted, so it is important that farmers contemplating retirement should discuss this in detail with their accountants.

2. Retirement relief

Retirement relief in respect of capital gains tax is given in respect of the sale of assets such as milk quota. The regulations in respect of this retirement relief are complex and to discuss this further is beyond the scope of this book. Attention, however, is drawn to its significance as it is most important that any dairy farmer contemplating retirement should discuss its implications with his accountant and other advisers.

3. Adequate income and somewhere to live

Having adequate income and somewhere to live are possibly the most important factors determining whether or not successful retirement is achieved.

Many farmers build up a successful business and then find they cannot transfer the business to successors without the need for a crippling rent equivalent from the successor in terms of consultancy fees, etc.

Reference was made earlier, in Chapter 14, to a farmer who has successfully passed on his business to the next generation. This has been made feasible in part by having an adequate pension fund as well as making appropriate decisions in the business during the past decade, including the appropriate training/provision of opportunities to his successor(s).

MAKE A WILL

Having thought very carefully about the above, it is important that both parents make a will so their wishes are implemented, otherwise the estate will be passed on by the 'laws of intestacy'. These will ensure the assets are passed on, but the way in which they are passed on will almost certainly not be in the way that was desired. It will almost certainly take a very long time and the successor(s) could have great difficulty in continuing with the business.

It is therefore vital that a will is made in good time. The ideal time never arrives. Hopefully, persons planning to retire will live for a long time and make changes to the will as time passes.

These days people generally are living longer and in many instances the grandchildren reach the age to start farming well before the grandparents' demise.

CHAPTER 18

Postscript: Selling and Marketing our Milk

MILK MARQUE

I started to write this book in October 1993 and expected that by the time it was finished the whole future of Milk Marque would have been settled, based on the expected vesting date of 1 April 1994. It is now 1 April 1994 and I am putting the finishing touches to the book.

During the past few days I have received a press release from the Ministry stating that the Milk Marketing Board is to be given a further two months in which to make its submissions on its revised proposals for Milk Marque. Presumably, this means that the Ministry are going to request even more changes and one feels there is a campaign to try to weaken the bargaining power of dairy farmers who, at the present time, are mainly resolved to join Milk Marque. I trust that this will not happen and that we shall see a strong Milk Marque as from 1 November 1994.*

Farmers, and that includes dairy farmers, are constantly urged to pay more attention to marketing. In the past our hands have been tied as we have been obliged to delegate the job of selling our milk to the Milk Marketing Board. We have *had* to leave them to get on with it. My view is that we shall have to do the same with Milk Marque, the difference being that in the future, we shall be able to vote with our feet if we are not satisfied with their performance and leave Milk Marque to join some other organisation.

* August 1994. This is proving to be the case.

However, the delay in the setting up of Milk Marque is cause for concern. Today's press, for example, carries headlines to the effect that there is talk of 60, yes 60, fledgling producer groups contemplating setting up a business in competition with Milk Marque, presumably on the assumption that they will be able to do the job better than the professionals.

In the same week the press carries reports of the keys to the success of one of the most successful independent milk marketing organisations in Northern Ireland:

1. It was stressed that the most important factor is to have tankers that can move the milk to the buyer. Milk Marque can do this.
2. You need a minimum of four buyers. Milk Marque has many more than four.
3. You need a buyer of milk you can depend on. Milk Marque, presumably, has Dairy Crest.
4. You need more than one financial backer or 'buyers will try to strangle you'. Milk Marque has been telling us this for a long time.
5. You need to use a computerised agency to manage administration: 'a producer group needs to be better than the big boys'. Milk Marque has also been telling us this for some considerable time.

To conclude, my view is that we need a strong Milk Marque. I trust that by the time this book is published, we will have an agreed vesting date for Milk Marque and that the majority of producers will decide to join it.

NICHE MARKET

Having stated my views that there is a need for a strong Milk Marque, I believe that there will also be niche markets and that some efficient producer/processing groups will eventually emerge with more than just a toe-hold in the marketplace. These are unlikely to include producer groups that sell to one processor.

These successful producer/processor groups are likely to have found the key to 'adding value' to an excellent raw material, ie 'high quality milk produced by the UK dairy farmer'.

Liquid milk sales will probably remain the best added-value product, although we are constantly being told that there is a massive growth potential in other value-added products, which at the present time is satisfied by buying imports from the rest of Europe.

Adding value to produce these products is certainly profitable at the present time, but one is bound to question the extent to which they will remain profitable as the GATT proposals are implemented. These include proposals to reduce the subsidies on exports, and to allow the imports of more milk products, eg from New Zealand.

Some 25% of the UK's butter is already sourced from New Zealand. UK processors are urged by the Ministry to do something to avoid this increase (eg by Mark Roche, ADAS Dairy Business Consultant at an AMC/ADAS Dairy Production and Economics Conference, in March 1994) but why should they if the added value they can receive in the market cannot compete with the New Zealand imports?

The Milk Marketing Board, urged on by the Ministry, invested producers' funds in the purchase of plants to manufacture butter back in the 1970s only to have to close these down in the 1980s following the introduction of milk quotas. We do not want to do that again.

The GATT provisions will also lead to a reduction in the level of subsidised export of cheese and will allow increased imports. It is suggested by some experts that these proposals could lead to an actual increase in the world market price, which in turn would lead to an increase in the price for cheese within the United Kingdom. This conclusion would, however, appear to be debatable.

Time will tell, and that is what we as producers need, ie time before we start to meddle in milk marketing. Many changes are likely to take place during the next few years which none of the experts to date have envisaged and some of these could be of an opportunist nature.

Entrepreneurs, including farmer-entrepreneurs, will see these opportunities, and I trust that in due course we shall see farmer-led ventures into the processing and distribution of milk.

CAPITAL REQUIREMENTS

Much is said about the very high capital requirements for the investments necessary in the processing of milk, £20–30 million being quoted as the amount required for a relatively small, but efficient plant. At first sight this seems a tremendous amount of capital until one looks at the amount of capital that dairy farmers already have invested in the production of milk.

A producer owning 250 acres and producing one million litres of milk has an investment in land and quota of approximately £1 million and a further investment of approximately £250,000 in dairy cows and youngstock, ie a total of £1.25 million. One hundred dairy farmers of this size therefore have a total investment of £125 million, five times the amount of capital required to set up an efficient processing factory.

To improve profits in the future, the same 100 farmers are probably contemplating a 25% increase in their production involving the purchase of both cows and quota, representing an investment in total of, say, £50 million.

We as producers at some stage need to step back and ask ourselves whether we are right to continue to compete amongst ourselves for the small share of the producers' market, or to take a decision to co-operate together and add value to at least part of our product.

This book in effect is about survival of the fittest. My gut feeling is that it would pay these 100 farmers to invest collectively £25 million in a processing plant, rather than to continue simply to increase production.

COST OF PRODUCING MILK

All the talk at the present time is about ways and means of receiving 1p or 2p extra for the milk that we sell, either to Milk Marque or whatever organisation we as producers decide to join. No mention is made of the cost of production in relation to these various sales outlets, and in this connection I am talking about the cost production in terms of seasonality, rather than in terms of efficiency and closeness to market.

A tremendous amount of pressure has been put on farmers over recent years to change their seasonality of production, the objective being the sacred cow of a level milk supply; yet no one, but no one receives a premium at the present time for producing a level supply. The recent changes in the seasonality prices for milk have led to an increase in the number of cows shown as being in-calf in the June returns for the United Kingdom, from 8% in 1988 to 13% in 1993.

This trend is likely to continue as more cows/heifers in the pipeline have already been served to calve at this time. The consequence of these changes, which are now difficult to reverse, is an almost certain shortfall in milk supplies during the winter months, or at least a shortfall in terms of an even supply.

The industry now has to get its act together to decide whether this is a good or bad thing. The cost of producing milk in the winter months is 5–6p less than producing it in the summer months. There is not likely to be a shortfall in relation to the supply of high added-value products, eg liquid cream and yogurt, but there is likely to be a shortfall for products with low added-value, particularly butter. If we are to meet the challenge from New Zealand, therefore, we need to look at the possibility of competing with them on even terms, by aiming to produce most of the milk for butter production during the summer months. In other words, producers and processors need to get together and look at the total cost of producing milk for butter, based on summer milk production.

Recent changes in the genetic potential of dairy cows, and the cost of alternative forages to silage, are leading producers to produce higher yields per cow based on farming systems in which the cows are housed for a high proportion of the year, with much less emphasis on grass production. This is not a very exciting prospect for farmers who have to base their milk production systems on grass, eg in the West and the North West, and to compete these farmers are going to have to increasingly concentrate on summer milk production to take advantage of their longer grazing season and generally good grass-growing conditions.

Farmers producing sugar beet have an autumn campaign to process their sugar beet into sugar. Why shouldn't the dairy farmers of the West and North have a summer campaign to convert their milk production into butter and cheese, with both producer and the processor closing down in the winter months?

PREMIUM TO EVEN-OUT IMBALANCE IN LOCALLY SOURCED SUPPLY

In recent weeks and months the author has attended various meetings and read various articles regarding the sale of milk to purchasers other than Milk Marque. In most instances, the emphasis has been on savings that can be made on haulage and administration and in virtually all cases the prices proposed have been based on those proposed by Milk Marque, including seasonality.

Quite a few of the potential purchasers intend to source milk locally as well as purchasing from Milk Marque and are offering local farmers prices based on those paid by Milk Marque.

Detailed questioning has revealed that this local supply will not meet one of the main needs of the local purchasers which is an even supply of milk and that they are expecting to *pay a premium* to Milk Marque to even out their sourced supply. This is the nub of the problem. Producers are being offered a price based on what Milk Marque is prepared to pay, not a price based on the needs of the purchasers.

One of the strengths of Milk Marque is that it will be able to 'ship' milk from areas such as the South West to areas that are in short supply during the winter months.

The only thing that one can say with confidence about the future of milk marketing is that changes will occur that up until now have not been envisaged. Opportunities will occur and will be developed by entrepreneurs and it will be in the light of these developments that producers in the future will have to make their decision whether or not to stay with their existing purchaser.

It would seem fair to say, however, that producers are not likely to change their purchasers of milk frequently as, in effect, each producer will be joining a 'club' and the success of this club will depend upon the loyalty between producers and purchasers.

Appendices

APPENDIX 1

Terms and Definitions Used in Farm Business Management

1 The terms defined below relate in the main to financial transactions covering a period of twelve months; generally this is the accounting year, but when dealing with crops it is sometimes the harvest year, although the calculations can be made for any period if required. For cash flow purposes shorter periods of time, eg, one month or one quarter are often used.

VALUATIONS

2 The process of valuation is essentially one of estimation. The basis to be used may vary according to the purpose for which the valuation is made and alternative bases are given in some instances in the following section. The basis of valuation used in any data presented should be clearly stated and should be consistent throughout the period of any series of figures.

(a) *Saleable crops in store* are valued at estimated market value including area payments less costs still to be incurred, eg, costs of marketing and storage. Market value and costs still to be incurred may be those either at the date of valuation or at the expected date of sale.

(b) *Growing crops* are valued at estimated cost up to the date of valuation. This may be either at variable costs or at estimated total cost. For most purposes, variable costs are preferable. Residual manurial values need to be taken into account only on change of occupancy.

Source: Ministry of Agriculture, Fisheries & Food.

(c) *Saleable crops ready for harvesting* but still in the ground should preferably be treated as 'Saleable crops in store' and should be valued as in 2(a) above less the estimated cost of harvesting. Alternatively, they may be treated as 'Growing crops'.

(d) *Fodder stocks (home-grown)* may be valued at estimated market value or variable costs. In calculating gross margins, valuation at variable costs is generally to be preferred. If market value is used, stock of non-saleable crops, eg, silage, should be valued in relation to hay value adjusted according to quality. Fodder crops still in the ground, eg, kale, turnips, are treated as growing crops.

(e) *Stocks of purchased materials (including fodder)* are valued at cost.

(f) *Machinery and equipment* are valued at original cost net of investment grants, less accumulated depreciation to date of valuation. Depreciation may be calculated by either the straight line or the reducing balance method.

(g) *Livestock,* whether for breeding, production or sale, are valued in their present condition at current market value, less cost of marketing. Fluctuations in market value which are expected to be temporary should be ignored.

OUTPUT TERMS

3 *Sales* are the value of goods sold for cash or on credit. In trading accounts sales exclude machinery and capital equipment. These items would, however, be included in capital accounts and cash flow calculations.

4 *Receipts* are monies received during the accounting period from the (see paragraph 3 above) sale of goods plus other remuneration, eg, subsidies, contracting, wayleaves. If practical, receipts should be recorded before deduction of off-farm marketing expenses such as commission and hire of containers.

5 *Revenue (or income)* is Receipts adjusted for debtors, including outstanding subsidies such as area payments, at the beginning and end of the accounting period.

6 *Returns* are Revenue adjusted for valuation changes.

Note: The value of goods and services produced on the farm for which no payment is made, eg, produce consumed in the farmhouse or supplied to workers, is not included in Returns but does form part of Gross Output.

7 *Gross Output* is total Returns plus the value of produce consumed in the farmhouse or supplied to workers for which no payment is made, less purchases of livestock, livestock products and other produce bought for resale. This can be calculated either for the farm as a whole or for separate sectors. Total Gross Output includes the Gross Output of livestock, crops and items of miscellaneous output such as certain subsidies and grants, contracting, wayleaves.

8 *Gross Output of Livestock* (as a whole or for individual enterprises) is total Returns from livestock and livestock products less purchases of these items plus the value of any livestock produce consumed in the farmhouse, or supplied to workers for which no payment is made. Gross Output of Livestock includes production grants attributable to the enterprise, eg, hill sheep subsidy, suckler cow subsidy.

9 *Gross Output of Crops* (as a whole or for individual enterprises) is total Returns from crops plus the value of any crop produce consumed in the farmhouse or supplied to workers for which no payment is made less the value of any produce bought for resale.

10 *Enterprise Output of a Livestock Enterprise** is its Gross Output plus the market value of livestock and livestock products transferred to another enterprise (transfers out) plus the market value of any production from the enterprise consumed on the farm less the market value of the livestock and livestock products transferred from another enterprise (transfers in).

11 *Enterprise Output of a Sale Crop Enterprise** is the total value of the crop produced, irrespective of its disposal; it equals Returns from the crop plus the market value of any part of the crop used on the farm. When this is calculated for the 'harvest year', as distinct from the accounting year, valuation changes are not relevant and the total yield of the crop is entered at market values plus deficiency payments.

12 *Enterprise Output from Forage* consists primarily of the sum of the Enterprise Outputs of grazing livestock. In addition it includes keep let and occasional sales eg, hay†, together with an adjustment for changes in the valuation of stocks of home grown fodder. *Note:* Changes in stocks caused by yield variations attributable to weather conditions, the severity or length of the winter or minor changes in livestock numbers or forage acres can be omitted from forage output and regarded as an item of miscellaneous output. Adjusted Enterprise Output from Forage is Enterprise Output from Forage less rented keep and purchase of bulk fodder.

13 *Net Output* is Gross Output less the cost of purchased feed, livestock keep, seed, bulbs, cuttings and plants for growing on.

14 *Standard Output* is the average Output (as defined in paragraphs 9–11) per acre of a crop or per head of livestock calculated as appropriate from national or local average price and yield data.

INPUT TERMS

15 *Purchases* are the value of materials and livestock acquired. In trading accounts purchases exclude machinery and capital equipment. These

* The Gross Output of an enterprise can be less than its Enterprise Output if any product of that enterprise is retained for use on the farm.

† Where sales of seed or fodder crops such as hay are a regular part of farm policy, they should be regarded as cash crops, not as forage crops.

items would, however, be included in capital accounts and cash flow calculations.

16 *Payments* are monies paid during the accounting period for purchases (see paragraph 15 above) of material, livestock, and for services, off-farm marketing expenses, other owner-occupier expenses, interest and loan repayment. Non-cash items such as unpaid labour and estimated rental value should not be included.

17 *Expenditure* is Payments adjusted for creditors at the beginning and end of the year.

18 *Costs* are Expenditure with the following adjustments:

Add:

(a) the opening valuation of the cost item;
(b) the depreciation on items of capital expenditure including machinery;
(c) to depreciation any loss made on the sale of machinery (*ie*, the difference between written down value and sale price);
(d) to the cash wages of workers the value of payments-in-kind (if not already included in the earnings figure used);

Deduct:

(a) the closing valuation of the cost item;
(b) purchases of livestock, livestock products and any other produce bought for resale;
(c) from depreciation any profit made on the sale of machinery (see c above);
(d) from machinery costs any allowance for the private use of farm vehicles;
(e) the value of purchased stores used in the farmhouse, (eg, coal, electricity), or sold off the farm, from the cost of that item.

19 *Inputs* are Costs with the following adjustments made in order to put all farms on a similar basis for comparative purposes:

Add:

(a) the value of unpaid family labour, including the manual labour of the farmer and his wife;
(b) for owner-occupiers and estimated rental value, less any cottage rents received and less an allowance for the rental value of the farmhouse;

Deduct:

(a) any mortgage payments and other expenses of owner-occupation;
(b) interest payments;
(c) cost of paid management;
(d) a proportion of the rental value of the farmhouse on tenanted farms.

20 *Fixed and Variable Costs for use in Gross Margin Calculations.* Variable Costs are defined as those costs which both can be readily allocated to a specific enterprise and will vary in approximately direct proportion to changes in the scale of that enterprise. The main Variable Costs are seed, fertilisers, sprays, concentrate feedingstuffs and much of the casual labour and contract machinery.

Fixed Costs are those costs which cannot readily be allocated to a specific enterprise and/or will not vary in direct proportion to small changes in the scale of the individual enterprises on the farm. Fixed Cost items include regular labour, machinery depreciation, rent and rates, and general overheads. Fuel and repairs are usually treated as Fixed Costs but glasshouse fuel is general treated as a Variable Cost.

MARGIN TERMS

21 *Management and Investment Income* is the difference between Gross Output and Inputs. It represents the reward to management and the return on tenant's capital invested in the farm, whether borrowed or not.

22 *Net Farm Income* is Management Investment Income less paid management plus the value of the manual labour of the farmer and his wife. It represents the return to the farmer and his wife for their own manual labour and their management and interest on all farming capital, excluding land and buildings.

23 *Profit* [*or Loss*] is the difference between Gross Output and Costs. It represents the surplus or deficit before imputing any notional charges such as rental value or unpaid labour. In the accounts of owner-occupiers it includes any profit accruing from the ownership of the land.

24 *Gross Margin of a Crop Enterprise* is its Enterprise Output less its Variable Costs.

25 *Gross Margin of a Livestock Enterprise* (*non-land using*) is its Enterprise Output less its Variable Costs. (For barley beef, variable costs include those of any hay fed).

26 *Gross Margin from Forage* is Enterprise Output from Forage less (see paragraph 12):
   (a) the Variable Costs directly attributable to individual types of grazing livestock, such as purchased concentrates, market value of home grown cereals fed, veterinary and medicine, AI fees;
   (b) the Variable Cost of forage and catch crops, such as fertiliser, seed, etc;
   (c) the cost of purchased forage and rented keep.

27 *Gross Margin of a Grazing Livestock Enterprise* (*excluding Forage*) is its Enterprise Output less its Variable Costs.

28 *Gross Margin of a Grazing Livestock Enterprise* (*including Forage*) is the Gross Margin of a grazing livestock enterprise (excluding Forage) less the allocated Variable Costs of forage, the cost of purchased forage and rented keep where such an allocation is considered to be both possible and useful.

29 *Other Margin Terms.* When other uses of the term margin are made, they should be fully described on the basis of the definitions above.

OTHER TERMS

30 *Tenant's Capital* [*or Operating Capital*] is the estimated amount of capital on the farm, other than land and fixed equipment. There is no easy way of determining this sum precisely and estimates are made in several ways depending on the information available and the purpose for which the estimate is required. One method is to take the average of the opening and closing valuations of stores (feed, seed, fertilisers), machinery, crops and livestock; alternatively the closing valuation only may be taken. A third estimate is obtainable by calculating the annual average of several estimated valuations during the year. Whichever of these methods is used, the valuation may be at cost or at market value. Any estimate produced should be accompanied by a description of the method of calculation.

APPENDIX 2

Tax Facts

INCOME TAX RELIEFS

	1993/94 £	*1994/95* £
Persons under age 65		
Personal	3,445	3,445
Married couple's	1,720	1,720*
Persons aged 65 but under 75		
Personal	4,200	4,200
Married couple's	2,465	2,665*
Persons aged 75 and above		
Personal	4,370	4,370
Married couple's	2,505	2,705*
Income limit for age allowance	14,200	14,200
Allowance reduced by 50% of excess (but not below level of main personal allowances)		
Additional relief for single parent	1,720	1,720
Widow's bereavement	1,720	1,720
Blind person	1,080	1,200
Enterprise investment relief limit	—	100,000*

* Married couple's allowance, enterprise investment relief and mortgage interest relief are restricted to relief at 20% for 1994/95.
Mortgage interest relief is restricted to interest on the first £30,000 of any mortgage.

TAX RATES

1994/95 Taxable Income	*Band*	*Rate*	*Tax on Band*	*Total Tax*
£	£	%	£	£
0–3,000	3,000	20	600	600
3,000–23,700	20,700	25	5,175	5,775
Excess		40		
1993/94				
0–2,500	2,500	20	500	500
2,501–23,700	21,200	25	5,300	5,800
Excess		40		

The tax credit on dividends is 20%, and also satisfies basic rate liability.

INHERITANCE TAX

Transfers made after 9 March 1992

Death rates		
Gross transfer	£	*Rate %*
First	150,000	Nil
Excess		40

There is taper relief on transfers more than three but not more than seven years before death. Chargeable lifetime transfers are initially charged at 20% instead of 40%.

Exemptions
There are exemptions for small gifts: £250 per donee. Annual gifts: £3,000 per donor. In consideration of marriage: parent £5,000; grandparent, remoter ancestor or party to marriage £2,500; other £1,000.

CORPORATION TAX

Financial year to	*31/3/94*	*31/3/95*
Full rate	33%	33%
Small companies rate	25%	25%
Small companies limit	£250,000	£300,000
Effective marginal rate	35%	35%
Upper marginal limit	£1,250,000	£1,500,000
Small companies fraction	1/50	1/50
Advance corporation tax	9/31	1/4

CAPITAL GAINS TAX

Rate	Taxed as top slice of income
Annual exempt amounts:	*£*
Individuals, personal representatives for year of death and two years thereafter and trusts for mentally disabled or those in receipt of attendance allowance, etc.	5,800
Other trusts generally	2,900

Retirement relief for capital gains

First £250,000 and one-half of gains between £250,000 and £1,000,000 exempt where business held for ten years. (For disposals after 18 March 1991 and before 30 November 1993, these figures were £150,000 and £600,000 respectively.) Minimum age for relief 55 (or earlier on grounds of ill health). Relief restricted where business held 1–10 years.

CAR BENEFIT RATES

For 1994/95 onwards, the cash benefit of a car provided by an employer is 35% of its 'price'. The benefit is reduced by one-third where business mileage exceeds 2,500 pa, and is reduced by two-thirds where business mileage exceeds 18,000 pa. Any resulting benefit is further reduced by one-third where the car is four or more years old at the end of the year of assessment. There are special provisions for 'classic cars', 'second cars' and employee capital contributions.

CAR FUEL SCALE RATES

	1993/94		*1994/95*	
	Petrol	*Diesel*	*Petrol*	*Diesel*
Cylinder capacity	£	£	£	£
Up to 1400cc	600	550	640	580
1401cc to 2000cc	760	550	810	580
2001cc or more	1,130	710	1,200	750

These scale charges are reduced to nil if the employee is required to make good the cost of all fuel provided for private use, and does so.
These figures also represent the annual VAT scale charge. The quarterly figures are one-quarter of the annual charge, and the monthly figures are one-twelfth of the annual charge.

CAPITAL ALLOWANCES

Machinery and Plant
First-year allowance 40% on expenditure incurred after 31 October 1992 and before 1 November 1993.
Writing-down allowance 25% (reducing balance basis).
Industrial and Agricultural Buildings
Initial allowance 20% on expenditure incurred after 31 October 1992 and before 1 November 1993, where building brought into use before 1 January 1995.
Writing-down allowance 4% (straight line basis).

VALUE ADDED TAX

Standard rate 17.5%
Registration level £45,000 per annum (previously £37,600)
Deregistration limit £43,000 per annum (previously £36,000)
Cash Accounting Scheme max. turnover £350,000 pa
Annual Accounting Scheme max. turnover £300,000 pa

NATIONAL INSURANCE CONTRIBUTIONS

1994/95

Class 1—Employees

Employees pay no contribution if their weekly earnings are below £57.00. Otherwise, employees are liable as follows:

Weekly earnings	*Not contracted out*	*Contracted out*
On first £57	2%	2%
On balance up to £430	10%	8.2%
Over £430 flat rate of max. pw	£38.44	£31.72

Men aged 65 or over, and women aged 60 or over, do not pay employees' contributions. However, employers' contributions are still payable.

Class 1—Employers	*Not contracted out*	*Contracted out*
Below £57.00	—	—
£57.00–£99.99	3.6%*	0.6%†
£100–£144.99	5.6%*	2.6%†
£145.00–199.99	7.6%*	4.6%†
£200–£430.00	10.2%*	7.2%†
Over £430	10.2%*	(Note 1)

* On all earnings

† Over £57, the first £57 is chargeable at the appropriate 'Not Contracted Out' rate.

(Note 1) £32.67 pw + 10.2% on earnings over £430 pw.

Class 1A—Employers: on car and car fuel

10.2% of the aggregate of the appropriate car benefit and car fuel scale rates (see pages 241–242).

Class 2—Self-employed Earnings (£3,200 or more)

Flat rate £5.65 pw.

Class 3—Voluntary

Flat rate £5.55 pw.

Class 4—Self-employed

7.3% of annual profits between £6,490 and £22,360 pa. Half contributions allowable for income tax purposes.

SOCIAL SECURITY BENEFITS

Weekly benefit	*1993/94* £	*1994/95* £
Basic retirement pension	56.10	57.60
Addition for wife or other adult dependant	33.70	34.50
Child benefit		
eldest child	10.00	10.20
subsequent children	8.10	8.25
Invalid care allowance	33.70	34.50
Increase for wife or other adult dependant	20.15	20.65
Unemployment benefit		
Single person	44.65	45.45
Wife or other adult dependant	27.55	28.05

RETIREMENT ANNUITIES AND PERSONAL PENSIONS

1994/95
Age at 6 April in year of assessment

Personal pensions (max. %)

35 and below	17.5%	51–55	30%
36–45	20%	56–60	35%
46–50	25%	61 and over	40%

No relief available for contributions on earnings exceeding £76,800 (previously £75,000 for 1993/94).

Retirement Annuities (max. %)

50 and below	17.5%	56–60	22.5%
51–55	20%	61 and over	27.5%

STATUTORY SICK PAY

From 6 April 1994

Lower rate	£47.80
Standard rate	£52.50

STATUTORY MATERNITY PAY

From 6 April 1994

Higher rate	9/10 of employee's average weekly earnings
Lower rate	£48.80 per week

APPENDIX 3

Herd Yield Prediction

The introduction of milk quotas has led to a great deal of interest in predicting what the milk production will be on a month-by-month basis, so that an individual farm finishes the year on quota.

The information shown in this appendix illustrates how a herd yield prediction can be used and the problems that are involved in ensuring that the budget herd yield is achieved.

In this example a herd yield prediction was made before the start of the quota year, ie on 5 March 1993. A second prediction was made in the light of progress on 16 July 1993 and a third on 22 October 1993. A further prediction was also made towards the end of the year to fine tune production against quota but this is not included in this data.

Actual production was monitored against the original prediction and finished the year at 1,210,807 litres compared to an original estimate of 1,292, 666 (see Table A3.1).

Production in April was in line with prediction but fell approximately 10,000 litres below prediction in May and this trend continued during the months June, July, August, September, October and November and was only arrested in December by taking a decision to purchase additional cows.

The revised herd yield prediction produced in October (Table A3.3) proved to be an accurate estimate of the likely income.

Milk quota had previously been leased in to produce 1.29 million litres and a decision was therefore taken in October to lease out the surplus quota, ie 80,000 litres.

There are many herd yield prediction computer programmes on the market. The advantage of the one illustrated is its simplicity and the ease with which the programme can be updated as the year progresses based on actual production.

The only information required to update the programme is:

Herd yield at the start of the month.
Number of cows expected to calve each month and peak yield.
Number of heifers expected to calve each month and peak yield.
Number of cull cows and number of cows dried off each month.

Table A3.1 Example farm (processed 5.03.1993)

Month		At start of month			Cows dried off and culled		Heifers		Cows		At end of month		Average yield per day				
		Daily yield	No. in milk	No. in herd	Dry	Cull	No. calving	Peak yield per day	No. calving	Peak yield per day	Daily yield	No. in milk	Total	No. in milk	Yield per cow	Yield in month	Cumulative yield
Apr	Actual		177	204		3			8							67938	
	Budget	2400	158	205	25	—			3	30	2240	136	2320	147	16	69600	69600
May	Actual		136	201		1			39							70918	
	Budget	2240	136	205	39	—			44	30	2985	141	2613	139	19	80988	150588
Jun	Actual		146	200		3	1		23							79894	218750
	Budget	2985	141	205	34	—			25	30	3131	132	3058	137	22	91733	242320
Jul	Actual		146	198		4			40							97068	
	Budget	3131	132	205	15	—			39	30	3852	156	3491	144	24	108236	350557
Aug	Actual		147	194		5			26							112408	
	Budget	3852	156	205	12	3			34	30	3982	175	3917	166	24	121433	471990
Sep	Actual		152	189		1			15							111121	539347
	Budget	3982	175	202	7	3			15	30	3944	180	3963	178	22	118886	590876
Oct	Actual		155	188		5			13							107794	
	Budget	3944	180	199	5	3			12	30	3837	184	3891	182	21	120608	711484
Nov	Actual		157	183		4	21		4							105012	
	Budget	3837	184	196	5	3	20	25	7	30	4092	203	3965	194	20	118936	830420
Dec	Actual		189	210		5			15							120771	872924
	Budget	4092	203	213		—			5	30	3832	208	3962	206	19	122824	953244
Jan	Actual		200	205		1	11		3							123962	
	Budget	3832	208	213		—	20	25	5	30	4099	233	3966	221	18	122942	1076186
Feb	Actual		203	215		—	4									110240	
	Budget	4099	233	233	2	—	5	25			3796	236	3948	235	17	110538	1186724
Mar	Actual		193	219		—	2					177				103681	1210807
	Budget	3796	236	238	42	—					3039	194	3417	215	16	105942	1292666
Year	Actual			200		32	38		186								
	Budget			210	186	12	45		189								

Table A3.2 Example farm (processed 16.07.1993)

Month		At start of month			Cows dried off and culled		Heifers		Cows		At end of month		Average yield per day				
		Daily yield	No. in milk	No. in herd	Dry	Cull	No. calving	Peak yield per day	No. calving	Peak yield per day	Daily yield	No. in milk	Total	No. in milk	Yield per cow	Yield in month	Cumulative yield
Apr	Actual			204		3			8							67938	
	Budget			204		3			8							67938	67938
May	Actual			201		1			39							70918	138856
	Budget			201		1			39							70918	138856
Jun	Actual			200		3			23							79894	218750
	Budget			200		3	1		23							79894	218750
Jul	Actual																
	Budget	3000	146	198	17				39	30	3717	168	3359	157	21	104114	322864
Aug	Actual																
	Budget	3717	168	198	16	3			30	30	3722	179	3719	174	21	115298	438162
Sep	Actual																
	Budget	3722	179	195	7	3			17	30	3769	186	3746	183	20	112366	550528
Oct	Actual																
	Budget	3769	186	192	3	3			15	30	3788	195	3779	191	20	117148	667676
Nov	Actual																
	Budget	3788	195	189	2	3	20	25	7	30	4075	217	3932	206	19	117947	785623
Dec	Actual																
	Budget	4075	217	206		3			5	30	3730	217	3902	217	18	120975	906598
Jan	Actual																
	Budget	3730	217	203			20	25	2	30	3917	239	3824	228	17	118534	1025132
Feb	Actual																
	Budget	3917	239	223	8		5	25			3578	236	3748	238	16	104939	1130071
Mar	Actual																
	Budget	3578	236	228	40						2861	196	3220	216	15	99805	1229876
Year	Actual																
	Budget			203	93		46										

Table A3.3 Example farm (processed 22.10.1993)

Month		At start of month			Cows dried off and culled		Heifers		Cows		At end of month		Average yield per day				
		Daily yield	No. in milk	No. in herd	Dry	Cull	No. calving	Peak yield per day	No. calving	Peak yield per day	Daily yield	No. in milk	Total	No. in milk	Yield per cow	Yield in month	Cumulative yield
Apr	Actual																
	Budget			204		3			8							67938	67938
May	Actual																
	Budget			201		1			39							70918	138856
Jun	Actual																
	Budget			200		3	1		23							79894	218750
Jul	Actual																
	Budget			198		4			40							97068	315818
Aug	Actual																
	Budget			194		5			26							112408	428226
Sep	Actual																
	Budget			189		1			15							111121	539347
Oct	Actual																
	Budget	3550	155	188					7	30	3405	162	3478	159	22	107803	647150
Nov	Actual																
	Budget	3405	162	188		4	29	27	10	30	4112	197	3758	180	21	116506	763656
Dec	Actual																
	Budget	4112	197	213		3			6	30	3853	200	3982	199	20	123455	887111
Jan	Actual																
	Budget	3853	200	210		13	19	25	3	30	3916	209	3885	205	19	120425	1007536
Feb	Actual																
	Budget	3916	209	216	8		5	25			3577	206	3747	208	18	104908	1112444
Mar	Actual																
	Budget	3577	206	221	40						2860	166	3219	186	17	99775	1212219
Year	Actual																
	Budget			202	48	37	54		177								

APPENDIX 4

Quotas

1. BASICS

Full details of the initial legislation and the subsequent changes to the arrangements have been given in previous editions of *Dairy Facts and Figures*. In this edition, only a summary of the system is presented together with details of any recent changes. It must be stressed that the application of the quota system differs in detail between the various countries of the Community and that the description given here relates only to the United Kingdom.

A system for a supplementary levy (*superlevy*) to be charged on milk produced above a specified reference quantity (*quota*) was introduced in the European Community in 1984 to run for a period of five years, the intention being to correct the imbalance between output and consumption within the Community. In 1988, the system was extended for a further three years, and in 1992 the existing arrangements were extended for a ninth year, up to 31 March 1993. From 1 April 1993, the system was extended for a further seven years (see **8**).

The quota system covers all sales of milk or milk products from farms. Two categories of quota were created: *dairy (wholesale) quota* corresponding to Deliveries to Dairies and *direct sales quota* corresponding to direct sales of milk and milk products by farmers to consumers. Each Member State was allocated a definitive wholesale and direct sales quota. When new Member States joined the Community, they have been given definitive quotas. Following the re-unification of Germany, an additional allocation of quota was made specifically for the territory of the former GDR. From time to time, Member States have been given permission to convert quota of one type into an equal amount of quota of the other type, usually from direct into wholesale. Until 1993, this is the only means by which these definitive quotas had changed since being first agreed.

The amount of this definitive quota which was actually available in any year was the subject of negotiation as part of the annual price agreement.

Reproduced with permission from UK Dairy Facts and Figures, *1993 edition, Federation of UK Milk Marketing Boards.*

An additional 1% was made available for the first year. Available quota was reduced by 2% in 1987–8 and another 1% in 1988–9. In 1987–8, there was also a 'temporary suspension' of 4% of wholesale quota, increasing to 5.5% in 1988–9. In 1989–90, available quota was cut by 1% but this was balanced by a reduction in the 'temporary suspension' from 5.5% to 4.5%. From April 1991, quota was cut by a further 2%. By 1992–3 therefore, 6% of the definitive quota had been cut and 4.5% suspended, leaving only 89.5% as available.

In addition to this 'normal' quota, there was a *Community reserve*. From this, an additional allocation of quota was made to the Republic of Ireland in view of the 'special significance of the dairy sector to Irish agriculture'. An equivalent amount was allocated to the UK specifically for Northern Ireland. Later, quota reserves were established for so-called **SLOM** cases and for producers with 'special needs' (commonly referred to as **Nallet** quota). These are discussed in **2**. The allocations of definitive and reserve quota to each Member State for 1992–3 are shown on Table A4.1.*

2. SLOM AND NALLET QUOTA

Those milk producers who participated in the Conversion and Non-Marketing of Milk Premium Scheme operated by the Community in the period 1977–1981 who, on 1 April 1984, were still within the prohibition period required under the terms of that Scheme and therefore not in milk production, did not qualify for a quota. The fact that the quota system prevented them from re-starting milk production at the end of their period of prohibition was judged by the European Court as unacceptable. Additional quota was therefore created from which such producers (referred to as SLOM producers) could receive an allocation.

In 1989, Ministers agreed the 'Nallet' package of measures which included the creation of a new Community reserve amounting to 1% of the definitive quota. Member States were free to allocate this quota to producers according to certain guidelines subject to Commission approval. The UK allocated this quota primarily to small producers.

3. LEVIES ON WHOLESALE SALES

The operation of the quota system is exceedingly complex; what follows aims to present the key features.

The UK operates '*Formula B*' under which liability for wholesale levy is assessed initially at the level of first purchaser which then recovers the money from its producers. Each of the Boards is a purchaser and there are also other bodies judged to be 'purchasers' for quota purposes. Assessment for levy involves comparing deliveries and quota in the period 1 April–31

* Tables are on pages 256ff.

March; in leap years, deliveries on leap-year day are not counted. (In some parts of the Community, assessment is on the basis of periods of 52 weeks for local administrative reasons.)

The purchaser is the sum of its individual producers. A purchaser's deliveries is the total of the 'deliveries', adjusted for butterfat (see **6**), of all its producers. Where a producer has transferred quota in or out during the year (see **7**), his 'deliveries' will be net of any production which was transferred with that quota.

Under the quota regulations in force up to 1992–3, those producers who held both direct sales quota and wholesale quota at the year-end were effectively treated as if they had a single quota: they were only considered as being over quota if their direct sales plus 'deliveries' exceed their total quota. To allow this principle to operate, it is frequently necessary to treat quota of one type as if it were quota of the other type. Take for example a producer whose direct sales fell short of his direct sales quota but whose 'deliveries' exceeded his wholesale quota: he can use his surplus ('unused') direct sales quota to reduce his excess 'deliveries'. The 'unused' portion of his direct sales quota is therefore converted into wholesale quota up to the point at which his wholesale quota now equates to his 'deliveries'. Such *year-end interchanges* can of course take place in either direction. Since a purchaser's quota is the aggregate of the wholesale quota held by all its producers, it follows that each such interchange results in a corresponding change to the purchaser's quota. The sum total of all quota (direct plus wholesale) at the national level remains unaffected by interchanges. It must be stressed that interchanges take place only for the purpose of calculating levy liability at the year-end: **no actual transfer of quota is involved!**

The quota rules allow that a Member State only pays levy if its national 'deliveries' exceed quota. Hence there will only be a levy in the UK if the aggregate position for all its purchasers (and hence for all producers, since purchasers are the aggregates of their producers) shows 'deliveries' to be in excess of quota. Levy will only be paid on any net UK surplus. To implement this principle, any surplus 'unused' wholesale quota has to be *re-allocated* between purchasers. This is done first within and then between regions. Thus, if a purchaser is under quota (after interchanges), its unused quota will be used to reduce (and possibly eliminate) the excess of those purchasers in the same region who are over quota. Any unused quota remaining is then used to reduce (and possibly eliminate) the remaining excess for purchasers in other regions.

Interchanges and re-allocations result in a significant difference between the actual wholesale quota held by a purchaser's producers and that effectively available to it when assessing levy liability as can be seen from Table A4.6. The effective wholesale quota for a purchaser (and hence its final levy position) cannot be determined until well after the quota year has ended.

The above points explain the difference between the final net position for levy purposes as shown on Table A4.6, and the figures shown on Table A4.4 which is how the position appears in simple statistical terms.

Two purchasers paid superlevy in 1985–6 (the England and Wales Board and the Northern Ireland Board) and in 1986–7 three purchasers (the England and Wales Board, the Scottish Board and that for Aberdeen and District) had to pay levy. In 1987–8, all except the North of Scotland Board paid levy and in 1988–9 for the first time all Boards paid levy. There was no wholesale levy in the UK in 1989–90. In 1990–91, no purchaser in Scotland paid levy but certain purchasers in other areas paid levy including the Boards in England and Wales and Northern Ireland. In 1991–2, only purchasers in Northern Ireland, together with the SMMB, paid levy. In 1992–3, there was no wholesale levy for any UK purchaser.

4. WHOLESALE LEVIES TO PRODUCERS

Through the complex maze of the quota system, two clear facts stand out. Firstly, *those producers whose deliveries remain within their quota can never become liable for levy*. Only where deliveries exceed quota is there any possibility of paying levy on this excess production (**potentially liable production**). Secondly, *where a purchaser has not had to pay levy, none of the producers delivering to that purchaser will be levied* even if their production has exceeded their quota (ie was potentially liable for levy). Only where a purchaser pays levy will it need to recover this from its over-quota producers.

Since 1989–90, the levy rate has been 115% of the Target Price prevailing on 31 March of that quota year. Under the current rules, only those producers most over quota (in percentage terms) pay levy. The point beyond which producers start to pay levy is termed the *threshold percentage*. Those producers with potentially liable production but who have exceeded their quota by less than the threshold percentage do not pay any levy. Those producers who exceed quota by more than the threshold are subject to levy but only on that part of their potentially liable production which exceeds the threshold: a producer who exceeds his quota by 10% when the threshold is 7% will pay levy on the 3% by which he has exceeded the threshold.

The threshold is calculated such that the total volume of potentially liable production which exceeds this level (and thus is actually subject to levy) equals the final net surplus deliveries on which the purchaser must pay levy. In this way, the purchaser recovers in full the total sum it has paid to the Intervention Board. Each purchaser will have its own threshold. For any purchaser, the larger the final net surplus on which it pays levy, the lower the threshold it has to apply.

5. LEVIES ON DIRECT SALES

In principle any producer who makes direct sales in excess of his direct sales quota would be liable for levy on that excess. However, interchanges as described in **3** will modify this situation. In addition, the rules allow that, provided total direct sales in the UK (after interchange) remain within the total UK direct sales quota, there is no direct sales levy. Because of this provision, there was no direct sales levy prior to 1988–9 and again in 1989–90.

The UK pays levy on the net excess at the direct sales levy rate of 75% of the Target Price (ie 20.3ppl at 31 March 1993) and then recovers this sum from individual producers. Any direct seller who exceeds their direct sales quota (after any interchange) is liable for levy on **ALL** their excess production. Since the total gross excess (liable) litres greatly exceeds the net excess on which the UK pays levy, the actual rate of levy on producers' excess direct sales is greatly diluted.

In 1988–9, the levy rate to producers was just under 1.4ppl; in 1990–1, it was approximately 4.2ppl; in 1991–2 it was almost 1.5ppl; in 1992–3 it was almost 5.6ppl.

6. BUTTERFAT

Since 1987–8, all wholesale quota has had a *butterfat reference figure* (*base*) associated with it. Each producer therefore has not only his quota but his butterfat base. Whenever quota moves from one producer to another, it takes with it the butterfat base of the producer giving up the quota; the base of the recipient is re-calculated as a weighted average of the additional quota and his existing quota base. Likewise, when production is transferred with quota (see **7**), the transferred litres take with them to the recipient the corresponding butterfat average of the donor producer's milk production.

At the year-end, the weighted average butterfat content of each producer's 'deliveries' for the year (including any deliveries transferred in) is calculated. The producer's 'deliveries' are then adjusted up (or down) by 0.18% for every 0.01 percentage point that this average differs from his base, to give his **butterfat-adjusted deliveries**. It is this figure which is then used in all levy-assessment calculations. Such an adjustment is only made when the average butterfat of 'deliveries' for all producers in the UK exceeds the average butterfat base of all these producers (as weighted by their quota). This condition has been fulfilled each year starting in 1989–90.

7. LEASING AND QUOTA TRANSFERS

Producers can alter the amount of quota they hold by either leasing or transfer between themselves and another producer.

Leasing is confined to wholesale quota and is very simple. All that is required is notification to the relevant authority, together with the fee to cover costs, before a certain deadline. Under UK Regulations, the deadline in 1992–3 was 30 November. A lease applies only for that quota year: the quota reverts to the original producer at the start of the next year. In 1992–3, as in earlier years, leasing could only take place between producers having the same purchaser and so did not affect the total quota held by a purchaser. Statistics on leasing are shown on Table A4.7.

By contrast, permanent transfer of quota can be a complicated legal matter and differences in the system of land tenure within the UK mean that there is no uniform legal procedure. However, there is now such a body of legal expertise that quota can be traded relatively easily with many thousands of transfers taking place each year. Transfers were only permitted within a region, the country being divided into 8 regions. This meant that in most cases transfers took place between producers having the same purchaser. When a transfer of quota takes place, a pro-rata share of the donor's production in that year up to the date of transfer is also transferred. A transfer therefore results in the 'deliveries' of both the donor and the recipient being different from their actual physical deliveries.

8. NEW QUOTA REGULATIONS

Continual amendment of the original 1984 EC Regulations led to the decision to re-draft them. Member State quotas as at 31 March 1993 were consolidated, bringing together all the changes made over the years, and incorporating the quota reserve. Subsequently, the further needs of SLOM producers were met by increasing wholesale quota by 0.6% for those Member States affected (including the UK), whilst substantial additions were made to the quotas given to Greece, Italy and Spain. The new quotas are set out on Table A4.2.

UK Regulations implementing the new EC Regulations made the following changes:

- apart from some small Scottish islands, the UK is now treated as a single region.
- the direct sales levy rate was made equal to that for wholesale sales at 115% of the Target Price.
- a new Permanent Conversion facility allows any producer to convert any amount of one type of quota into an equal amount of the other. Conversion is permanent but applications can only be made between 1 April and 31 December in any year.
- interchange (see **3**) is re-named Temporary Conversion and there is no longer a requirement to hold both types of quota.
- the leasing deadline may be any date up to 31 December.
- producers who neither lease out or produce against their quota in a quota year will have their quota confiscated and placed in the national reserve. It may be reclaimed if the producer returns to milk production within 6 years.

As a consequence of virtually all of the country now being treated as one region, regional balancing of unused quota for the purpose of the levy calculation (see **3**) will no longer apply; balancing will take place on an equal basis between all purchasers, unused quota being allocated to over-quota purchasers in direct proportion to their contribution to the UK excess. (Note: it is likely that new UK Regulations from 1994–5 will allocate unused quota to over-quota purchasers in proportion to the purchaser's **total** quota.) Unallocated quota held in the national reserve will participate in this balancing. Wholesale quota may now be transferred almost anywhere in the UK.

The introduction of Permanent Conversion means that the total amount of quota of any type is no longer fixed; only the total of wholesale and direct sales quota for a Member State is fixed. This means that the figures given in Table A4.2 will be amended once applications have been processed. Direct sales quota may now, effectively, be leased by first converting it into wholesale quota using Permanent Conversion.

These changes increase the uncertainty in predicting a purchaser's effective wholesale quota prior to the year-end.

Table A4.1 Milk quotas 1992–3 by Member State

Member State	Wholesale quota								Direct sales quota
	Definitive guaranteed quantity	*Guaranteed quantity for year*	*Temporary suspension*[b]	*Available reference quantity*	*Reserve quota*	*SLOM quota*	*'Nallet' quota*	*Total*	
	A	*B*	*C*	*D=B–C*	*E*	*F*	*G*	*H=D+E+F+G*	
	tonnes[c]								
Belgium	3,211,000	3,025,531	144,495	2,881,036	—	6,574	32,110	2,919,720	373,193
Denmark	4,882,000	4,589,080	219,690	4,369,390	—	9,620	48,820	4,427,830	951
France	25,634,000	24,195,960	1,153,530	23,042,430	—	64,027	256,340	23,362,797	732,824
Germany[a]	30,227,000	28,514,420	1,360,215	27,154,205	—	161,046	234,230	27,549,481	150,038
Greece	537,000	544,780	24,165	520,615	—	—	5,370	525,985	4,528
Ireland	5,280,000	4,963,200	237,600	4,725,600	303,000	117,958	52,800	5,199,358	15,210
Italy	8,798,000	8,620,120	395,910	8,224,210	—	—	87,980	8,312,190	717,870
Luxembourg	265,000	249,100	11,925	237,175	25,000	1,674	2,650	266,499	951
Netherlands	11,979,000	11,248,260	539,055	10,709,205	—	47,886	119,790	10,876,881	102,307
Portugal	1,779,000	1,743,420	—	1,743,420	—	—	—	1,743,420	118,580
Spain	4,650,000	4,571,000	209,250	4,361,750	50,000	—	46,500	4,458,250	516,950
UK	15,329,574	14,392,824	689,831	13,702,993	65,000	191,215	153,296	14,112,504	392,868
Total	112,571,574	106,657,695	4,985,666	101,672,029	443,000	600,000	1,039,886	103,754,915	3,126,270

[a] Includes GDR.
[b] 4.5% of Definitive guaranteed quantity as shown in A, except Portugal (exempt).
[c] Figures rounded to nearest tonne where necessary.

Table A4.2 Milk quotas 1993–4 by Member State

	Wholesale quota					
Member State	*1992–3 quota*[a]	*SLOM*[c]	*Other awards*	*Converted from direct sales quota*	*1993–4 quota*[f]	*Direct sales quota*[f]
	tonnes					
Belgium	2,919,720	17,518			2,937,238	373,193
Denmark	4,427,830	26,567			4,454,397	951
France	23,362,797	140,177			23,502,974	732,824
Germany[b]	27,549,481	165,297		+ 50,000	27,764,778	100,038
Greece	525,985		100,000		625,985	4,528
Ireland	5,199,358	31,196			5,230,554	15,210
Italy	8,312,190		900,000		9,212,190	717,870
Luxembourg	266,499	1,599			268,098	951
Netherlands	10,876,881	65,261	30,243[d]	– 281	10,972,104	102,588
Portugal	1,743,420		10,461[e]	+ 51,000	1,804,881	67,580
Spain	4,458,250		591,750	+150,000	5,200,000	366,950
UK	14,112,504	84,675			14,197,179	392,868
Total	103,754,915				106,170,378	2,875,551

[a] See Table A4.1.
[b] Includes former GDR.
[c] A further 0.6% added to meet latest round of SLOM claims.
[d] Technical adjustment to convert 52-week figure to 365-day equivalent: this does not affect the effective amount of quota available to Dutch producers.
[e] Equivalent to 0.6% SLOM award although SLOM did not apply to Portugal.
[f] As at 14 June 1993; these figures will be subject to revision as a result of Permanent Conversion by individual producers—see text.

Table A4.3 United Kingdom wholesale quota: allocation between regions

Region		*1990–1*[a]	*1991–2*[b]	*1990–1*	*1991–2*
		kilograms		*million litres*[c]	
1	**England and Wales** (including Scilly Isles)	11,832,511,254	11,611,833,155	11,491.262	11,276.948
2	**Northern Ireland**				
	Basic	1,256,552,496	1,238,306,511	1,220.314	1,202.594
	Special Allocation[d]	65,000,000	65,000,000	63.125	63.125
	Total	1,321,552,496	1,303,306,511	1,283.439	1,265.719
	Scottish MMB area				
3	All areas other than those in 4	990,796,432	970,883,931	962.222	942.884
4	Islands and Kintyre	57,830,473	56,890,212	56.163	55.249
	Total	1,048,626,905	1,027,774,143	1,018.385	998.133
5	**Aberdeen MMB area**	108,261,439	107,806,994	105.139	104.698
	North of Scotland MMB area				
6	All areas excluding Orkney	40,355,023	40,545,002	39.191	39.376
7	Orkney	15,634,190	17,047,617	15.183	16.556
	Total	55,989,213	57,592,619	54.374	55.932
8	**Shetland**	2,354,454	2,303,275	2.287	2.237
	Total Scotland	1,215,232,011	1,195,477,031	1,180.185	1,160.999
	Total United Kingdom	143,369,295,761	14,110,616,697	13,954.885	13,703.667

Figures for 1992–3 had not been published by MAFF at the time of going to press.
[a] As published in the London Gazette 10 April 1992.
[b] As published in the London Gazette 20 April 1993.
[c] Converted at 0.97116 kg per litre.
[d] From Community reserve.

Table A4.4 Comparison of deliveries and dairy quota: United Kingdom by Board area[a]

April to March	*England and Wales*	*Scottish MMB*	*Aberdeen and District*	*North of Scotland*	*Board Areas in Scotland*	*Northern Ireland*	*United Kingdom*
	million litres						
1989–90							
Dairy quota	11,515	1,018	105	54	1,178	1,286	13,978
Deliveries[c]	11,537	1,017	106	55	1,179	1,310	14,025
1990–1							
Dairy quota	11,493	1,018	105	54	1,177	1,283	13,954
Deliveries[c]	11,458	1,012	106	56	1,174	1,285	13,918
1991–2							
Dairy quota	11,278	998	105	56	1,159	1,266	13,703
Deliveries[bc]	11,157	1,009	104	56	1,169	1,278	13,604
1992–3							
Diary quota	11,278	998	105	56	1,159	1,266	13,703
Deliveries[c]	11,081	979	101	54	1,134	1,270	13,485

Users are strongly advised to read the accompanything text and the footnotes to Table A4.6 in order to appreciate the relationship between this table and Tables A4.3 and A4.8 and thus avoid misleading conclusions.

[a] Includes all deliveries to purchasers within areas covered by MM Schemes but excludes deliveries outside these geographical areas.

[b] Adjusted for leap year.

[c] Excluding butterfat adjustment.

Table A4.5 Number of producers over and under wholesale quota: United Kingdom by Board

Percentage over or under quota	*England and Wales*		*Scottish MMB*		*Aberdeen and District*		*North of Scotland*		*Northern Ireland*		*United Kingdom*	
	over	*under*	*over*	*under*	*over*	*under*	*over*	*under*	*over*	*under*	*over*	*under*
1991–2												
0–1.99[a]	5,256	4,187	460	415	44	22	11	13	1,115	1,277	} 15,023	10,690
2.00–4.99[b]	6,599	3,556	438	295	24	21	15	17	1,061	887		
5.00–9.99	3,892	2,544	259	158	9	18	5	20	624	439	4,789	3,179
10.00–19.99	824	1,632	68	84	4	8	2	10	236	237	1,134	1,971
20.00 and over	161	1,184	9	212	3	15	1	4	75	648	249	2,063
No production	—	2,657	—	156	—	—	—	4	—	599	—	3,416
No quota	28	—	1	—	—	—	—	—	4	—	33	—
Total %	51.5	48.5	48.3	51.7	50.0	50.0	33.3	66.7	43.3	56.7	49.9	50.1
1992–3												
0–1.99[a]	4,400	3,341	273	259	13	19	13	6	1,375	1,407	} 13,930	9,618
2.00–4.99[b]	6,646	3,131	326	320	23	20	5	11	856	1,104		
5.00–9.99	4,460	2,658	254	298	14	14	11	16	240	515	4,979	3,501
10.00–19.99	682	1,982	88	210	12	13	11	10	37	238	830	2,453
20.00 and over	86	1,404	31	298	8	26	4	10	6	815	135	2,553
No production	—	2,891	—	169	—	—	—	4	—	505	—	3,569
No quota	15	—	1	—	—	—	—	—	—	—	16	—
Total %	51.4	48.6	38.5	61.5	43.2	56.8	43.6	56.4	35.4	64.6	47.8	52.2

Figures are only for those producers consigning their milk to the Boards.
[a] Aberdeen Board band = 0.00–2.49.
[b] Aberdeen Board band = 2.50–4.99.

Table A4.6 Superlevy liability 1992–3: United Kingdom by area

	England and Wales[a]	Northern Ireland	Scottish MMB	Aberdeen and District	North of Scotland	Scottish Islands	United Kingdom
April to March				*million litres*			
Wholesale							
Quota (WSQ)	11,278	1,266	998	105	56	2	13,705
Sales	11,081	1,270	979	101	54	3	13,488
Adjusted sales[b]	11,307	1,285	984	101	55	2	13,734
Adjusted sales *less* Quota	+29	+18	–14	– 3	– 1	—	+29
Direct[c]							
Quota (DSQ)	315	20	24	10	2	2	374
Sales	305	6	18	9	2	2	343
Sales *less* Quota	–10	–14	– 6	– 1	—	—	–31
Net interchange (DSQ to WSQ)	+21	+14	+ 4	—	+ 1	—	+40
Post-interchange sales *less* Quota:							
Wholesale	+ 8	+ 4	–18	– 3	– 2	—	–11[d]
Direct	+11	0	– 2	– 1	+ 1	—	+ 9[e]

Users are strongly advised to read the accompanying text in order to appreciate the relationship between this table and Tables A.43 and A4.4.
Figures relate to ALL milk produced and marketed in the areas shown: they are not confined to milk sold to the Boards.
[a] Includes Isles of Scilly.
[b] After adjustment for butterfat content.
[c] Based on quota returns received within the deadline.
[d] Wholesale sales within quota for UK as a whole therefore no wholesale levy for any purchaser.
[e] Direct sales exceeded quota for the UK as a whole resulting in a levy on every UK producer who, after interchange, had exceeded his direct sales quota. The levy of 5.6ppl applied to all excess litres.

Table A4.7 Quota leasing 1992–3[a]: United Kingdom by country

April to March	*England and Wales*	*Scotland*	*Northern Ireland*	*United Kingdom*
Total transactions	16,254	609	644	17,507
Quantity of milk involved (million litres)	730.8	32.6	18.2	781.6
Percent of quota	6.5	2.8	1.6	5.8
Producers leasing out quota	5,808[b]	267[c]	312	6,387
Producers leasing in quota	10,452[b]	483[c]	479	11,414

[a] As administered by the Milk Marketing Boards.
[b] Includes 347 who leased in and leased out.
[c] Includes 6 who leased in and leased out.

Index

Index

Numbers in italics indicate tables

S

T

V

W